T0214009

Pathways in Mathematics

Series Editors

Takayuki Hibi, Department of Pure and Applied Mathematics, Osaka University, Suita, Osaka, Japan

Wolfgang König, Weierstraß-Institut, Berlin, Germany

Johannes Zimmer, Fakultät für Mathematik, Technische Universität München, Garching, Germany

Each "Pathways in Mathematics" book offers a roadmap to a currently well developing mathematical research field and is a first-hand information and inspiration for further study, aimed both at students and researchers. It is written in an educational style, i.e., in a way that is accessible for advanced undergraduate and graduate students. It also serves as an introduction to and survey of the field for researchers who want to be quickly informed about the state of the art. The point of departure is typically a bachelor/masters level background, from which the reader is expeditiously guided to the frontiers. This is achieved by focusing on ideas and concepts underlying the development of the subject while keeping technicalities to a minimum. Each volume contains an extensive annotated bibliography as well as a discussion of open problems and future research directions as recommendations for starting new projects. Titles from this series are indexed by Scopus.

More information about this series at http://www.springer.com/series/15133

Elijah Liflyand

Harmonic Analysis on the Real Line

A Path in the Theory

 Birkhäuser

Elijah Liflyand
Department of Mathematics
Bar-Ilan University
Ramat-Gan, Israel

Regional Mathematical Center of Southern
Federal University
Rostov-on-Don, Russia

ISSN 2367-3451   ISSN 2367-346X   (electronic)
Pathways in Mathematics
ISBN 978-3-030-81894-4   ISBN 978-3-030-81892-0   (eBook)
https://doi.org/10.1007/978-3-030-81892-0

Mathematics Subject Classification: 42AXX, 26AXX

This book is published under the imprint Birkhäuser, www.birkhauser-science.com, by the registered
company Springer Nature Switzerland AG.
The registered company address is: Gewerbestrasse 11, 6330 Cham, Switzerland

To Ola and Michelle

Contents

Chapter 1
Introduction

Can this book be called a textbook? To a certain extent the answer is yes, but not exactly. It is, so to say, a schematic textbook. More precisely, what is presented is a summary of my experience after reading textbooks, monographs and basic papers as well as working in mathematics. The key word is in the second part of the title:"a path". This means that the main objective of these notes is very modest: to present a collection of basic, well-known and some lesser-known results that may serve as a background for future research in certain topics of harmonic analysis on the real line. Many of these results are also a necessary basis for multivariate extensions. The latter can be found in [34]: the present notes are an outgrowth of the first chapter of that book and much of the first part was generalized to the multivariate case in the second part of that book. Anticipating a natural question, I would like to say that the subject has by no means been exhausted, and there are enough open problems both in dimension one and, even more, in several dimensions. Honestly, the text before you is more an essay than a textbook, though it can be used as a skeleton for constructing certain special courses, and if not completely then for outlining the instructive matter.

1.1 Motivation and Background

This collection is motivated by the following problems. On the one hand, the Fourier series may be divergent at quite many points, which leads to applying summability. A summability method is usually defined by a sequence of multipliers. A special though rather general situation appears when this sequence is generated by a single function, more precisely, by the values of such a function on a certain lattice. It turns out that approximate and summability behavior of the corresponding means strongly relies on the behavior (integrability properties) of the Fourier transform of the generating function. This is one source for my interest in delicate properties of

© The Author(s), under exclusive license to Springer Nature Switzerland AG 2021
E. Liflyand, *Harmonic Analysis on the Real Line*, Pathways in Mathematics,
https://doi.org/10.1007/978-3-030-81892-0_1

the Fourier transform and related content. On the other hand, a one hundred-year old problem whether a trigonometric series is the Fourier series of an integrable function was a source for many results on the behavior of the Fourier transform as counterparts of those for series. In such problems, especially when the transformed function is of bounded variation, this issue is in many respects related to the other important transform, the Hilbert transform, and, correspondingly, to the real Hardy space. It turned out that by applying these tools, the author frequently faced the situation that certain facts are difficult or impossible to find in the literature, or some are well known but exist as folklore only.

In fact, several parts of the well-known book by Butzer and Nessel [10] cover some of these topics in full detail. Unfortunately, only some topics, and in fact the book and the above mentioned parts were written for a different general purpose. There are a few other books with useful information on certain relevant issues, to mention some, those by Titchmarsh [54], Bochner [9], Bary [4], Zygmund [57], or the series of books by Stein and Shakarchi. Many other good books in the list are either essentially multidimensional   or much more general. However, in everyday research certain extracts are always needed; here is my attempt to collect them step by step. I believe that many of us have a similar collection, a toolkit for daily needs. Maybe the closest book with which the present text can be compared is Goldberg's book [15]. In his review of that book, Rudin characterizes it as the one that "contains an elementary exposition of the basic theorems of Fourier analysis on the real line... There are also references to some very recent papers". He does not use the word "path" but that brief book can definitely be called the author's path to his activities in harmonic analysis on the real line. The latter is outlined in the appendix to that book; the same role is played by the last chapter of the present book.

1.2   Structure

A brief outline of the book is in order. In Chap. 2, certain commonly used spaces and related notions are introduced. More precisely, we briefly discuss continuous functions, including moduli of continuity and smoothness; Lebesgue spaces; absolutely continuous functions, and functions with bounded variation. Discussion on some other spaces, for example, Hardy spaces, is postponed to the parts where the corresponding machinery is prepared. In Chap. 3, certain aspects of the theory of trigonometric series are discussed. In addition to certain standard issues, more attention is paid to the difference between the Fourier series and trigonometric series. In Chap. 4, the Fourier transform is similarly introduced: not every detail but some features related to other issues of the chosen path. Chapter 5 naturally adds the Hilbert transform to this treatment. Many aspects of this operator are examined in detail, with numerous examples. In Chap. 6, the Hilbert transform is related to Hardy spaces, for which other characterizations are also discussed. In Chap. 7, certain aspects of Hardy inequalities are considered. On the one hand, the chosen items are less standard, while on the other hand, discussion through the author's new results is opened. This chapter is a kind of transition to the next one, where already discussed and new aspects of harmonic analysis on the real line are illustrated via the author's results, most of which are rather recent. In fact, Chap. 8 needs a special description since it differs from the other chapters. On the one hand, since each of its sections is devoted to a recent problem studied by the author, one cannot think of it as a part of the textbook. On the other hand, formulations and proofs are thoroughly tied to certain items of the previous chapters. Moreover, some sections give a more detailed view of the issues only mentioned before, the first of which concerns interpolation. In general, this chapter shows where and how the chosen matter is applied and, in our opinion, strengthens the theory with relevant applications. There is also a different explanation. It is known that Erdös used to ask colleagues "What is your theorem?" This book ends with partial answers to this question but mainly presents some tools I am using in my research.

Not every statement, especially in the chapters before the last two, is followed by a proof. For many of them, just a standard or easily accessible reference will be given. For some, the proof will be just outlined. Complete proofs will be given if they are brief enough, or, sometimes, if they are dear to the author or are of interest to him in a special way, say, seem to be instructive or just unusual. So to say, landmarks are preferred to exact navigation. In fact, the decision whether to present a proof or not is mostly due to the author's taste and maybe the role that this particular proof played in the author's interests or research. One more advantage of that random choice of the presented proofs is that there is no need for special exercises. There are plenty of results for which the reader can try to find a standard or one's own new proof.

Mostly, notation will appear in appropriate parts of the book. Some of it is collected at the end of the text. However, something very general is worth being said immediately. We shall denote by C absolute constants, they may be different in different occurrences. Let $a \lesssim b$ mean the inequality $a \leq Cb$ when we do not wish to take care of precise constants. For example, if an inequality is considered on a class of functions, the corresponding constant is first of all independent of the functions in question. By $a \asymp b$, we denote a two-sided estimate $C_1 b \leq a \leq C_2 b$ with C_1, C_2 being two constants, independent of essential parameters. If we wish to emphasize the dependence of a constant on certain parameters, they will be given as subscripts, say, like C_p, $C_{p,q}$, etc.

1.3 Before Reading the Book

One of the goals of the present notes is to show how the author's scientific interests rather than certain pedagogical tasks form a course. Hopefully, it also will be instructive, quite vast and far-reaching. However, there was an additional circumstance for writing this book. In October 2017, the author gave a mini-course in Fourier Analysis in the Pontificia Universidad Javeriana, Bogotá, Colombia, within the scope of the Week of Israel in this university. This experience added much to the understanding of certain features of the topic in their presentation to a disparate audience in the mini-course format and of the need for a collection to be written in an organized form. The author acknowledges not only the opportunity to participate in that event but also and mainly the hospitality of the hosts and friendship with the colleagues from that university and the Israeli embassy. Of course, prior to this event many other people influenced my knowledge and understanding of mathematics. A long list of such people can be found in the foreword to [34]. This list can be repeated or maybe even extended. The name and influence of my teacher Roald Trigub must be emphasized detachedly.

However, some people have played a special role in the preparation of this manuscript. My friends and colleagues Alex Iosevich, Andrei Lerner, Anatoly Podkorytov and Sergey Tikhonov offered me invaluable advice on the structure and certain concrete issues of the book. Miriam Beller's editing assistance in the preparation of this manuscript was invaluable. The editors of Birkhäuser Thomas Hempfling and Dorothy Mazlum always were patient, helpful and encouraging as this manuscript progressed to press. My very special and warmest thanks to these people.

This book represents a particular path, one that I chose to take through the wonderful labyrinth of mathematical analysis. Many others took different paths that resulted in different adventures and ideas. Nevertheless, the fundamentals of the journey described in this book lay at the heart of modern analysis and allow the reader to develop the skills needed to study a variety of areas of research. In short, the paths may be different, but the compass is essentially the same. Though this book was written from the personal perspective of a mathematician working in the topics outlined therein, it is not an automathography, as Halmos called his memoir [19] but it does contain certain elements of this term. I have gone along a specified path throughout my life focusing on the details on the path but also gazing around. The results of this journey is before you, the reader.

Chapter 2
Classes of Functions

In this chapter, we briefly present well-known classes of functions that will play an important role in our considerations. Of course, the list of such classes goes far beyond those discussed in this chapter. The latter classes will be overviewed (briefly) just because they are very general and commonly used. More specific classes will appear later in our story. In other words, if the main subject of the book is claimed to be prerequisites in harmonic analysis applicable to certain problems, this chapter is a kind of preliminaries to those prerequisites. Maybe the section about functions of bounded variation is special and it is therefore presented in more detail. After all, this topic has a special place in the author's preference scale. So to say, all spaces are equal, but some spaces are more equal than others, at least, in the eyes of the author.

2.1 Continuous Functions and Lebesgue Spaces

We certainly assume that the reader is aware of the notion of continuity and the main features related to it. Here we are going to present more specific notions connected to continuity and introduce some smoothness classes (see, e.g., [53] or [55]).

2.1.1 Continuous Functions

One of the main measures of smoothness properties are the so-called *moduli of smoothness*. One of them, the main one historically, is called the modulus of

© The Author(s), under exclusive license to Springer Nature Switzerland AG 2021
E. Liflyand, *Harmonic Analysis on the Real Line*, Pathways in Mathematics,
https://doi.org/10.1007/978-3-030-81892-0_2

continuity. Let a bounded function f be defined on $[a, b]$. For any $h > 0$, we define

$$\omega(f; h) = \sup_{\substack{t_1, t_2 \in [a,b], \\ |t_1 - t_2| \leq h}} |f(t_1) - f(t_2)|. \tag{2.1}$$

The following properties of the *modulus of continuity* are almost obvious.

1. If f is continuous, then $\lim_{h \to 0} \omega(f; h) = 0$.
2. Moduli of continuity are semi-additive:

$$\omega(f; h_1 + h_2) \leq \omega(f; h_1) + \omega(f; h_2).$$

3. $\omega(f; h)$ is monotone increasing.
4. For m a positive integer, there holds

$$\omega(f; mh) \leq m\omega(f; h);$$

if $\lambda > 0$ is not necessarily an integer, then

$$\omega(f; \lambda h) \leq (\lambda + 1)\omega(f; h).$$

5. For f_1 and f_2 defined on the same interval, we have

$$\omega(\lambda f_1 + \mu f_2; h) \leq |\lambda|\omega(f_1; h) + |\mu|\omega(f_2; h).$$

In fact, any function satisfying the properties (1)–(4) is called a *modulus of continuity type* function (or just modulus of continuity), regardless of f; it is the modulus of continuity of itself and is denoted by $\omega(h)$. It is frequently said that a function f belongs to the ω smoothness class if $\omega(f; h) \lesssim \omega(h)$.

We say that the function f belongs to the Lip α class (Lipschitz or Hölder class), if $\omega(f; h) \leq Ch^\alpha$. This is meaningful for $0 < \alpha \leq 1$, and if $\alpha > 1$ such a function is constant. Very often, saying that a function belongs to the Lipschitz class without indicating any specific α means that the case where $\alpha = 1$ is understood.

If instead of the sup-norm in (2.1) we consider the L^p norm, $1 \leq p < \infty$, the corresponding quantity is called the integral modulus of continuity (or p-modulus of continuity); to wit

$$\omega(f; h)_p = \sup_{0 < s \leq h} \left(\int_a^{b-s} |f(t+s) - f(t)|^p \, dt \right)^{\frac{1}{p}}. \tag{2.2}$$

We mention that if f is 2π-periodic on $\mathbb{R}$, then the integral is calculated over the period $[-\pi, \pi)$. This modulus possesses the same properties as that in (2.1). We can define the Lip α class in integral metrics in the same manner, it will be denoted by $\text{Lip}_p \alpha$.

If a function is defined on a larger interval, we can define its modulus of continuity on a smaller interval with a milder assumption that all the involved values of the function lie in that larger interval. This is the case for periodic functions.

If in these definitions instead of $f(t + s) - f(t)$ or the like, we take higher differences of the function, say of k-th order,

$$\Delta^k(f; s) = \Delta\left(\Delta^{k-1}(f; s)\right),$$

where $\Delta(f(\cdot); s) := \Delta^1(f(\cdot); s) = f(\cdot + s) - f(\cdot)$, the corresponding value is called the *modulus of smoothness*. It is denoted by $\omega_k(f; h)_p$.

2.1.2 Lebesgue Spaces

In harmonic analysis, it is frequently of importance what kind of integrability is used. We shall mostly apply the Lebesgue integral, though sometimes Riemann integrability will suffice. Therefore, the spaces involved will often be or related to the Lebesgue L^p spaces (see (2.2)). It is worth mentioning that in [38, Chapter 4, §6], the theory of Lebesgue integration is separately considered on $\mathbb{R}$. The norm of the L^p function f defined on a measurable set $D \subseteq \mathbb{R}$ is

$$\|f\|_{L^p(D)} = \|f\|_p := \left(\int_D |f(t)|^p \, dt\right)^{\frac{1}{p}},$$

if $1 \leq p < \infty$, and

$$\|f\|_\infty := \operatorname*{ess\,sup}_{t \in D} |f(t)|,$$

for $p = \infty$, where ess sup as usual means the same as the supremum over D but $t \in D$ rather almost everywhere, i.e., except on a set of measure zero. In symbols, for a function g, it can be written as

$$\operatorname*{ess\,sup}_{t \in D} g(t) = \inf\{C : g \leq C \text{ almost everywhere on } D\}.$$

It is worth noting (see [38, Chapter 4, §4.4]) that in the definition of the essential supremum, the lower boundary can be replaced by the minimum, so that we have $g \leq \operatorname*{ess\,sup}_D g$ almost everywhere. Indeed, if $\operatorname*{ess\,sup}_D g = +\infty$, this is obvious, and if $\operatorname*{ess\,sup}_D g = C_0 < +\infty$, then $g \leq C_0 + \frac{1}{n}$ almost everywhere for every natural n, and the desired assertion follows by passing to the limit.

It is of importance that the L^p norm of f is the same for $|f|$ and for the *decreasing rearrangement* of f. The latter notion is defined as follows. We first define the distribution function. Given a measure μ and a μ-measurable set $D \subset \mathbb{R}$, we call

$$\mu_f(s) = \mu(x \in D : |f(x)| > s), \quad s \geq 0, \tag{2.3}$$

the *distribution function* of the measurable function f. The decreasing rearrangement of f is the function $f^* : [0, \infty) \to [0, \infty]$ defined by

$$f^*(t) = \inf\{s \geq 0 : \mu_f(s) \leq t\}, \tag{2.4}$$

where we use the convention that the infimum for the empty set is infinity. One of the main properties of the decreasing rearrangements is

$$\int_D |f(t)|^p \, dt = \int_0^\infty f^*(t)^p \, dt.$$

The same notion and property are true for sequences. In fact, the decreasing rearrangement $d^* = \{d_n^*\}$ of a sequence $d = \{d_k\}$ is a permutation of this sequence in the decreasing (non-increasing) order. A more formal definition is given by

$$d_n^* := \inf\{m > 0 : |\{k \in \mathbb{N} : d_k \geq m\}| < n\}.$$

Here $|E|$ means the measure of the set E, which in case of a discrete finite set merely means the number of points in E. Among standard and well-known properties of these spaces let us discuss the *completeness* of the L^p spaces, where a wide range of the tools of real analysis is applied.

Theorem 2.5 *The space $L^p(D)$, $1 \leq p \leq \infty$, where D is the whole $\mathbb{R}$ or its measurable subset, is complete.*

Proof By definition, we have to prove that every fundamental sequence in this space converges to one of its elements. The case of L^∞ is simple, and we omit its proof. Let $\{f_n\}$ be a fundamental (Cauchy) sequence in $L^p(D)$, $1 \leq p < \infty$, that is,

$$\|f_n - f_m\|_{L^p(D)} \to 0 \quad \text{as} \quad m, n \to \infty.$$

Then there is an increasing sequence $\{n_k\}$ such that

$$\|f_{n_{k+1}} - f_{n_k}\|_{L^p}^p = \int_D |f_{n_{k+1}}(t) - f_{n_k}(t)|^p \, dt < \frac{1}{2^{kp}},$$

for $k = 1, 2, \ldots$. Hence the m-th partial sums S_m of the series

$$|f_{n_1}(t)| + \sum_{k=2}^{\infty} |f_{n_k}(t) - f_{n_{k+1}}(t)| \tag{2.6}$$

increase and their integrals

$$\left(\int_D |S_m(t)|^p \, dt \right)^{\frac{1}{p}} = \left(\int_D |f_{n_1}(t)|^p \, dt \right)^{\frac{1}{p}} + \sum_{k=2}^{m} \left(\int_D |f_{n_k}(t) - f_{n_{k-1}}(t)|^p \, dt \right)^{\frac{1}{p}}$$

are uniformly bounded by

$$\sum_{k=1}^{\infty} \frac{1}{2^k} = 1.$$

By the Beppo Levi theorem, the sequence $\{S_m\}$ and, correspondingly, the series (2.6) converges almost everywhere. Therefore the series

$$f_{n_1}(t) + \sum_{k=2}^{\infty} [f_{n_k}(t) - f_{n_{k-1}}(t)], \tag{2.7}$$

for which f_{n_k} are its partial sums, converges almost everywhere to a function

$$f(t) = \lim_{k \to \infty} f_{n_k}(t).$$

We thus have proved that any fundamental sequence in $L^p(D)$ contains an almost everywhere convergent subsequence. Let us demonstrate that this subsequence converges to the same function in L^p norm. Since the initial sequence f_n is fundamental, for any $\varepsilon > 0$ and k and l large enough, there holds

$$\int_D |f_{n_k}(t) - f_{n_l}(t)|^p \, dt < \varepsilon^p.$$

Hence we have a sequence of non-negative functions converging almost everywhere as $l \to \infty$. By the Fatou theorem, we can take the limit as $l \to \infty$ behind the integral sign in this inequality, which leads to

$$\int_D |f_{n_k}(t) - f(t)|^p \, dt \leq \varepsilon^p.$$

This implies both $f \in L^p(D)$ and $f_{n_k} \to f$ in L^p norm. Since $\{f_n\}$ is a Cauchy sequence, the existence of a subsequence converging to a limit implies that the sequence itself converges to the same limit, that is, $f_n \to f$ in L^p norm, exactly as required.   □

The famous Hölder and Minkowski inequalities are known for various spaces but mainly for the Lebesgue spaces and the corresponding sequence counterparts. For $p = 2$, Hölder's inequality is called by various combinations of the names Bunyakovskii, Cauchy and Schwarz. We give the following generalized Minkowski inequality:

Let D and E be measurable subsets of $\mathbb{R}$ and let $f(t, x)$ be a measurable function defined on $D \times E$, with $t \in D$ and $x \in E$; for $p \geq 1$. We have

$$\left(\int_E \left(\int_D |f(t, x)| \, dt \right)^p dx \right)^{\frac{1}{p}} \leq \int_D \left(\int_E |f(t, x)|^p \, dx \right)^{\frac{1}{p}} dt. \tag{2.8}$$

Such a matter, except maybe (2.8), is discussed in almost every book on real analysis; for a comprehensive reference, [22, Chapter VI] can be recommended, especially Theorem 202 and (6.13.9) therein. Somewhat more general results of this type can be found in [38, Appendix 13.6.5]. It is worth mentioning that (2.8) holds true not only for integrals but also for sums and the mixtures of sums and integrals.

In fact, many of the above notions have counterparts for sequences; for instance, see the definition of the corresponding spaces in (4.28).

One more notion important for integrable functions is the notion of a *Lebesgue point*. For a function $f \in L^1(\mathbb{R})$, the Lebesgue point x is defined by

$$\lim_{\delta \to 0} \frac{1}{\delta} \int_x^{x+\delta} |f(u) - f(x)| \, du = 0. \tag{2.9}$$

The main feature of these points is that for an integrable function, almost every point is a Lebesgue point, that is, (2.9) holds for almost every x (see, e.g., [38, Chapter 4, §4.9.2]).

An essential result due to Lebesgue and closely connected with the introduced notion of Lebesgue's points is the Lebesgue differentiation theorem.

Theorem 2.10 *For each $f \in \mathbb{R}$, we have*

$$\lim_{\delta \to 0} \int_{x-\delta}^{x+\delta} f(t) \, dt = f(x),$$

for almost every $x \in \mathbb{R}$.

2.1.3 The Hardy–Littlewood Maximal Function

One of the very important tools in harmonic analysis is the Hardy–Littlewood maximal function. It has several versions and a huge number of applications. We will present one of the standard variants, give only some basics and illustrate its applications by rather important one in Chap. 5. In our presentation, we shall follow [12].

The Hardy–Littlewood maximal function of a locally integrable function f on $\mathbb{R}$ is defined by

$$Mf(x) = \sup_{r>0} \frac{1}{2r} \int_{|t-x|<r} |f(t-x)|\, dt. \tag{2.11}$$

The value $+\infty$ is not excluded here. Again, this is one of the most important operators in harmonic analysis. First of all, we must know whether and where this operator is of *strong type* and of *weak type*. The former means that the operator is bounded taking L^p into L^p (of course, other spaces may be considered). The latter, or more precisely that the operator $A : L^p \to L^q, q < \infty$, is of weak q type, means that

$$|\{x \in \mathbb{R} : |Af(x)| > \lambda\}| \le \left(\frac{C\|f\|_{L^p}}{\lambda}\right)^q;$$

the weak type (p, ∞) means that A is a bounded operator from L^p to L^∞. We hope that there will not be any confusion of similar notations $|\cdot|$ for the Lebesgue measure if "$\cdot$" means a set and the absolute value if "$\cdot$" means a number.

Theorem 2.12 *The operator M is of weak type $(1, 1)$ and of strong type (p, p), $1 < p \le \infty$.*

Proof That the operator is bounded taking L^∞ into L^∞ immediately follows from the definition. If we prove the weak type $(1, 1)$ assertion, the rest will follow from the following celebrated Marcinkiewicz interpolation theorem.

Theorem 2.13 *Let A be a sublinear operator from $L^{p_0} + L^{p_1}$, with $1 \le p_0 < p_1 \le \infty$, that is weak (p_0, p_0) and strong (p_1, p_1). Then A is strong (p, p) for $p_0 < p < p_1$.*

Now, we proceed to the proof of the weak type $(1, 1)$ assertion. Let $E_\lambda = \{x \in \mathbb{R} : |Mf(x)| > \lambda\}$. If $x \in E_\lambda$, then there is an interval $|t - x| < r$ such that

$$\frac{1}{2r} \int_{|t-x|<r} |f(t)|\, dt > \lambda. \tag{2.14}$$

Let $K \subset E_\lambda$ be a compact set. Then $K \subset \bigcup I_x$, where I_x are various intervals centered at x. It follows from the well-known covering lemma that only a finite number of such intervals $\{I_j\}$ can be taken so that $K \subset \bigcup_j I_j$ and $\sum_j \chi_{I_j} \leq 2$, where χ_E is the indicator (characteristic) function of the set E. Therefore,

$$|K| \leq \sum_j |I_j| \leq \sum_j \frac{1}{\lambda} \int_{I_j} |f(t)| \, dt \leq \frac{1}{\lambda} \int_{\mathbb{R}} \sum_j \chi_{I_j} |f(t)| \, dt \leq \frac{2}{\lambda} \|f\|_{L^1}.$$

The second inequality follows from (2.14). The fact that the last inequality holds for every compact $K \subset E_\lambda$ completes the proof.   □

In what follows, there will be further indications of the use of interpolation methods, without details. More will be given in Sect. 8.1 of Chap. 8.

In order to make Theorem 2.12 even more meaningful, it is worth mentioning that for $f \in L^1(\mathbb{R})$, with f non-trivial, we always have $Mf \notin L^1(\mathbb{R})$. Indeed, there exists $R > 0$ such that

$$\int_{|t|<R} |f(t)| \, dt \geq \nu > 0.$$

Now, for $|x| > R$, there holds $\{t : |t| < R\} \subset \{|t - x| < 2|x|\}$. Therefore,

$$Mf(x) \geq \frac{1}{2|x|} \int_{|t|<R} |f(t)| \, dt \geq \frac{\nu}{2|x|},$$

where the right-hand side is by no means integrable.

2.1.4   Calderón-Zygmund Decomposition

One of the most effective tools to treat integrable functions is the Calderón-Zygmund decomposition. Here we adopt the multivariate presentation in [12, Chapter 2, §5] to the case of the real line. First, dyadic intervals are introduced. We denote by $\mathcal{I}_0$ the collection of intervals $[m, m + 1)$, $m \in \mathbb{Z}$. Obtaining the dyadic intervals by dilating these intervals by 2^{-k}, we denote this collection by $\mathcal{I}_k$. It is simply $2^{-k}\mathbb{Z}$ and possesses the following properties.

1. For any $t \in \mathbb{R}$, there is a unique interval in each $\mathcal{I}_k$ which contains t.
2. Any two dyadic intervals are either disjoint or one is entirely contained in the other.
3. A dyadic interval in $\mathcal{I}_k$ is contained in a unique interval of each family $\mathcal{I}_j$, if $j < k$, and contains two dyadic intervals of $\mathcal{I}_{k+1}$.

With this in hand, we will construct a discrete analog of an approximation of the identity (cf. Sect. 4.5 of Chap. 4). For a locally integrable function f, we set

$$E_k f(x) = \sum_{I \in \mathcal{I}_k} \left(\frac{1}{|I|} \int_I f(t)\, dt \right) \chi_I(x).$$

In probability language, $E_k f$ is the *conditional expectation* of f with respect to the σ-algebra generated by $\mathcal{I}_k$. We mention the fundamental property of this quantity:
   If J is the union of intervals in $\mathcal{I}_k$, then

$$\int_J E_k f(x)\, dx = \int_J f(x)\, dx.$$

Theorem 2.15 *Let f be an integrable non-negative function and let λ be a positive number. Then there exists a sequence $\{I_j\}$ of disjoint dyadic intervals such that*

1. $f(t) \le \lambda$ for almost every $t \notin \bigcup_j I_j$;

2. $\left| \bigcup_j I_j \right| \le \frac{1}{\lambda} \|f\|_{L^1}$;

3. $\lambda < \frac{1}{|I_j|} \int_{I_j} f(t)\, dt \le 2\lambda$.

Proof We first construct the sets

$$J_k = \{ x \in \mathbb{R} : E_k f(x) > \lambda \quad \text{and} \quad E_j f(x) \le \lambda \quad \text{if} \quad j < k \}.$$

In other words, $x \in J_k$ if $E_k f(x)$ is the first conditional expectation of f which is greater than λ. Indeed, for each integrable f, we have $E_k f(x) \to 0$ as $k \to -\infty$, which ensures the existence of such k. Now, we decompose each J_k into disjoint dyadic intervals contained in $\mathcal{I}_k$. All these intervals form the claimed family $\{I_j\}$. Let $t \notin \bigcup_j I_j$, then for every k, we have $E_k f(t) \le \lambda$. It is known (see [12, Chapter 2, Theorem 2.10 (2)]) that almost everywhere

$$\lim_{k \to \infty} E_k f(x) = f(x).$$

Therefore, we get $f(t) \le \lambda$ for almost every t. Part 2 is simple. We have

$$\left| \bigcup_j I_j \right| \le \sum_j \frac{1}{\lambda} \int_{I_j} E_k f(x)\, dx$$

$$= \frac{1}{\lambda} \sum_j \int_{I_j} f(x)\, dx = \frac{1}{\lambda} \|f\|_{L^1}.$$

The left inequality in the third property of the sequence follows from the definition of the sets J_k. Finally, if $\widetilde{I}_j$ is the dyadic interval containing I_k whose length is doubled, then the average of f over $\widetilde{I}_j$ is at most λ. By this,

$$\frac{1}{|I_j|} \int_{I_j} f(t)\, dt \le \frac{|\widetilde{I}_j|}{|I_j|} \frac{1}{|\widetilde{I}_j|} \int_{\widetilde{I}_j} f(t)\, dt \le 2\lambda.$$

This completes the proof.   □

In Chap. 5, while treating the Hilbert transform, we will see how the Calderón-Zygmund decomposition comes into play.

2.1.5 Absolute Continuity

The notion of absolute continuity is in order. Plainly speaking, we shall pose such an assumption on considered functions for the purpose of being able to restore a function from its derivative when integrating by parts.

Definition 2.16 A function f defined on an interval $[a, b]$ is called absolutely continuous on it if for every positive number ε, there is a positive number δ such that whenever a finite sequence of pairwise disjoint sub-intervals (x_k, y_k) of $[a, b]$ with $x_k, y_k \in [a, b]$ satisfies

$$\sum_k (y_k - x_k) < \delta$$

then

$$\sum_k |f(y_k) - f(x_k)| < \varepsilon.$$

The collection of all absolutely continuous functions on $[a, b]$ is denoted by $AC([a, b])$. There is an equivalent definition of absolute continuity. If one of them is chosen to be the definition, the other can be derived from it as a property of absolutely continuous functions.

Definition 2.17 A function f defined on an interval $[a, b]$ is called absolutely continuous on it if it can be written in the form

$$f(x) = f(c) + \int_c^x h(t)\, dt, \qquad x \in [a, b],$$

where $c \in [a, b]$ and h is Lebesgue integrable on $[a, b]$.

The notion of absolute continuity can be defined in a similar way in the case where the interval is not closed; however, the function h may be not integrable on it but only *locally* integrable, that is, integrable on every *closed subinterval*. The latter notion is meaningful if a function is defined on a set of infinite measure, say on $\mathbb{R}$ or on the half-axis; the first part of [34] is mostly devoted to such functions.

The following Lebesgue theorem presents the substantial property of absolutely continuous functions.

Theorem 2.18 *If f is a function that is absolutely continuous on an interval $[a, b]$, then it is differentiable almost everywhere, its derivative is integrable, and*

$$f(y) - f(x) = \int_x^y f'(t)\, dt$$

for any $x, y \in [a, b]$.

Comparing the above with the cases where the claim of absolute continuity is relaxed, it is worth mentioning that if we know only that f is an increasing function, then its derivative is measurable and

$$\int_x^y f'(t)\, dt \le f(y) - f(x),$$

which proves that the derivative is Lebesgue integrable (see, e.g., [40, Theorem 5, Chapter VIII, §2]).

2.2 Functions of Bounded Variation

In this section, we present the main information on functions of bounded variation. The book [34] gives an idea of the role of such functions in harmonic analysis. This notion goes back to Jordan. The idea of variation is closely related to the problem of rectifiability of curves. Also, it shows the way how to control discontinuities (of first kind, jumps). However, we will go to a different source for motivation of our interest in this class of functions. There is no doubt about the importance of the class of monotone functions is. However, this class is not closed with respect to the simplest algebraic operations. The class of functions of bounded variation does not have this disadvantage and is a proper extension of monotone functions. We also mention recent attempts to generalize monotonicity in different ways; see the survey type paper [165] and recent results of that kind in such a classical area as number series in [166].

We begin with basic facts for functions on a finite interval $[a, b]$. We follow some of the sources where this class is studied in an appropriate and/or interesting form, see, e.g., [38, 40, 52]. Moreover, even a whole book on this subject has recently appeared [3]. Of course, this book is recommended for a deeper study of the subject but to get the basics in a capsulized form other sources might be more convenient.

Consider an arbitrary partition τ of $[a, b]$ by means of the points $t_0 = a < t_1 < \ldots < t_m = b$ and, for a function f defined on $[a, b]$, set

$$S_\tau = \sum_{k=0}^{m-1} |f(t_{k+1}) - f(t_k)|.$$

Definition 2.19 The value

$$\sup_\tau S_\tau$$

is called the *total variation* of the function f on the interval $[a, b]$ and is denoted by $Vf := V_{[a,b]}f$. If $V_{[a,b]}f$ is finite, an alternative notation is $f \in BV([a, b])$ (or another set in place of $[a, b]$), f is called a *function of bounded variation*.

It is clear that by adding new partition points, S_τ may only increase. Let us list additional properties of the total variation and of functions with bounded variation.

(i) For any α, β, with $a \le \alpha < \beta \le b$, there holds

$$V_{[a,b]}f \ge |f(\beta) - f(\alpha)|.$$

(ii) A monotone bounded function f is of bounded variation, with

$$V_{[a,b]}f = |f(b) - f(a)|.$$

(iii) A linear combination of functions with bounded variation is again a function with bounded variation, in particular,

$$V_{[a,b]}(f + g) \le V_{[a,b]}f + V_{[a,b]}g$$

and

$$V_{[a,b]}(\alpha f) = |\alpha| V_{[a,b]}f, \quad \alpha \in \mathbb{R}.$$

(iv) Each function of bounded variation is bounded.

Proof The property immediately follows from (i).   □

A less obvious property of the total variation is additivity.

(v) If $a < c < b$, then

$$V_{[a,b]}f = V_{[a,c]}f + V_{[c,b]}f.$$

Proof If we consider the partitions τ of $[a, b]$ containing c, then obviously

$$S_\tau \le V_{[a,c]} f + V_{[c,b]} f.$$

If a partition does not contain c, we add c to get a new partition, which along with the preceding arguments gives

$$V_{[a,b]} f \le V_{[a,c]} f + V_{[c,b]} f.$$

On the other hand, the union of any two partitions τ' and τ'' of $[a, c]$ and $[c, b]$, respectively, is only a part of all possible partitions of $[a, b]$. Therefore, we have

$$S_{\tau'} + S_{\tau''} \le V_{[a,b]} f.$$

Going to the least upper bound first with respect to τ' and then with respect to τ'', we obtain

$$V_{[a,c]} f + V_{[c,b]} f \le V_{[a,b]} f.$$

Combining the two opposite inequalities yields the needed property.   □

This property applies both to the case of bounded and unbounded variation.

As one can see from (ii) and (iii), the difference of two increasing functions is a function of bounded variation. It follows from the last property that the converse assertion is also true.

(vi) A function is of bounded variation if and only if it can be represented as the difference of increasing functions (a particular case of Jordan decomposition for measures).

In addition, a function of bounded variation can be written as the difference of increasing functions that are continuous at the same points as f is.

All these are for real-valued functions; for a complex-valued function, these assertions hold for the real and imaginary parts.

Proof The function $f_1(t) = V_{[a,t]} f$ obviously increases. Such is the function $f_2(t) = V_{[a,t]} f - f(t)$ as well. Therefore, $f = f_1 - f_2$ is the desired representation. This and the continuity assertion readily follow from (i).   □

(vii) The product of two functions of bounded variation is again a function of bounded variation; the quotient of two functions of bounded variation is again a function of bounded variation provided that the denominator is bounded away from zero.

Proof The property of the product follows from the obvious relation

$$f(t_{k+1}) g(t_{k+1}) - f(t_k) g(t_k) = [f(t_{k+1}) - f(t_k)] g(t_{k+1})$$
$$+ f(t_k) [g(t_{k+1}) - g(t_k)]$$

and property (iii). The property for the quotient follows from

$$\frac{f(t_{k+1})}{g(t_{k+1})} - \frac{f(t_k)}{g(t_k)} = \frac{[f(t_{k+1}) - f(t_k)]\,g(t_k) - f(t_k)\,[g(t_{k+1}) - g(t_k)]}{g(t_{k+1})g(t_k)},$$

property (iii) and from that the denominator is bounded away from zero.   □

Of course, the boundedness does not mean the continuity. However, a function of bounded variation does not, in a sense, vary enormously from continuous functions. Property (vi) implies, from the known properties of monotone functions, that, in particular, the set of points of discontinuity of a function of bounded variation is at most countable.

Theorem 2.20 *Every function* $f \in BV([a;b])$ *has at most countably many discontinuities in the interval* $[a, b]$.

Proof By virtue of (vi), it is enough to prove the statement for the case where f is a nondecreasing function. Let D be the set of all discontinuity points of f. For each $x \in D$, choose an arbitrary rational $q(x)$ in the interval $(f(x-); f(x+))$. Since f is nondecreasing, it follows that $q(x) \le q(y)$ whenever $x, y \in D$ and $x \le y$. Hence, the mapping q provides a one-to-one correspondence between D and a subset of rational numbers. This proves that D is at most countable.   □

Connections between functions of bounded variation and absolutely continuous functions began to be studied almost a century ago, see, e.g., [180]. Today these connections are known in full. First, absolutely continuous functions and functions of bounded variation are related in the following way (for a short and clear proof, see [38, Chapter 4, §11, Theorem 11.3]).

Theorem 2.21 *If a function* f *is absolutely continuous on* $[a, b]$, *then it is of bounded variation, with*

$$V_{[a,b]}f = \int_a^b |f'(t)|\,dt.$$

Note that a function f absolutely continuous on $[a, b]$ can be written as the difference of absolutely continuous increasing functions.

We now continue to list the properties of functions with bounded variation. The structure of a function with bounded variation is revealed by Lebesgue's decomposition.

(viii) Every function of bounded variation f can be represented as

$$f(t) = f_{jump}(t) + f_{AC}(t) + f_{sing}(t), \tag{2.22}$$

where f_{jump} is the function of the jumps of f, f_{AC} is an absolutely continuous function and f_{sing} is a *singular function*. The latter means that f_{sing} is a non-constant function and its derivative vanishes almost everywhere. Of course, if f is a continuous function, f_{jump} does not exist in (2.22) as well as both f_{jump} and f_{sing} are absent if f is absolutely continuous.

Let us now make a few remarks on functions of bounded variation on infinite intervals. Good sources where certain peculiarities of this instance are studied are [47] or [10]. The total variations $V_{[a,\infty]}f$ and $V_{[-\infty,b]}f$ can be defined by

$$V_{[a,\infty)}f = \lim_{b\to\infty} V_{[a,b]}f$$

and

$$V_{(-\infty,b]}f = \lim_{a\to-\infty} V_{[a,b]}f.$$

It is frequently natural to deal with functions f not only of bounded variation but vanishing at infinity,

$$\lim_{|t|\to\infty} f(t) = 0,$$

written $f \in BV_0(\mathbb{R})$.

We shall use the standard notation

$$\|f\|_{BV} = \int_{\mathbb{R}} |df(t)|, \tag{2.23}$$

where we use the so-called Stieltjes integration.

There is one more interesting property of functions of bounded variation that allows one to understand them as functions with certain integral smoothness. Indeed, on the one hand, for f of bounded variation and $h \in \mathbb{R}$, we obtain the inequality

$$\int_{\mathbb{R}} |f(t+h) - f(t)| \, dt \le \int_{\mathbb{R}} \left| \int_t^{t+h} df(s) \right| dt \le \|f\|_{BV} |h|,$$

that is, $\omega(f; h)_1 = O(h)$, or, in other words, this function belongs to the Lip 1 class in L^1. On the other hand, this property characterizes, in a sense, functions of bounded variation. More precisely, if $\omega(f; h)_1 = O(h)$, then almost everywhere f coincides with a function of bounded variation; [116] is probably the earliest source where this fact is stated.

Very often not only a single function of bounded variation is studied but a sequence of such functions. In such cases, Helly's theorems are of crucial importance (see, e.g., [40, Chapter VIII]). We formulate two of them.

Theorem 2.24 *If we have an infinite family of functions, defined on $[a, b]$, such that all the functions of this family as well as their total variations are bounded by one positive number, then there exist a sequence of functions in this family which converges at each point of $[a, b]$ to a function of bounded variation.*

Helly's second theorem is concerned with the passage to the limit under integral sign.

Theorem 2.25 *Let we have an infinite family of functions* $\{g_n\}$, *defined on* $[a, b]$ *and converging at each point of* $[a, b]$ *to a finite function* g, *and a continuous function* f *defined on the same segment. If the total variations of all the functions of this sequence are bounded by one positive number, then*

$$\lim_{n \to \infty} \int_a^b f(t) dg_n(t) = \int_a^b f(t) dg(t).$$

If the interval $[a, b]$ is infinite, then certain additional conditions may be needed. This is discussed in Theorem 4.38 (see Sect. 4.5.2 of Chap. 4).

Contrary to the notation $BV_0(\mathbb{R})$, that is, the space of BV functions vanishing at both $+\infty$ and $-\infty$, we will denote the space of BV functions vanishing at $-\infty$ regardless of their behavior at $+\infty$, by BV_-, though the norm will be kept as $\|\cdot\|_{BV}$. In fact, BV_0 is the set of BV_- functions for which

$$\int_{\mathbb{R}} df(t) = 0.$$

For $g \in L^1$, the function f defined by

$$f(t) = \int_{-\infty}^t g(s) \, ds$$

is an element of BV_-, and thus $AC(\mathbb{R}) \subset BV_-$.

This also allows one to proceed to integration by parts with respect to a measure generated by a function of bounded variation, As in the "usual" case, there is the Riemann-Stieltjes integration and more general Lebesgue-Stieltjes integration.

The formula for the Lebesgue-Stieltjes integration by parts reads as follows:

$$\int_{a+}^{a+b} f(x) \, dg(x) = (fg)(a + b) - (fg)(a) - \int_{a+}^{a+b} g(x-) \, df(x). \quad (2.26)$$

Of course, $g(x-)$ denotes the left limit of g at x. Standard assumptions on f and g might both be of bounded variation on every finite interval of $\mathbb{R}$ (see, e.g., [23, 21.67 and 21.68]). Of course, under an additional assumption, say, absolute continuity, there might be appropriate changes in this formula.

Much interesting information can be found in Chapter VIII of [40] in the Riemann-Stieltjes setting. For example, the integral

$$\int_a^b f(t) \, dg(t) \quad (2.27)$$

exists if f is continuous and g is of bounded variation. If both functions are only of bounded variation, one should be careful if there is a point of discontinuity common for both functions; otherwise no problems appear. It is also worth mentioning an interesting sufficient condition for the existence of (2.27) due to Kondurar and given as an exercise to that chapter: one of the functions satisfies the Lipschitz condition of order α, while the other of order β, and $\alpha + \beta > 1$.

Remark 2.28 In general, Lebesgue-Stieltjes integrals are usually considered with respect to arbitrary Borel measures rather than for functions of bounded variation. But there is a close connection between Borel measures and BV-classes. Indeed, apart from a different normalization in the interior points (instead of

$$f(t) = \frac{f(t+0) + f(t-0)}{2}$$

one assumes f to be left-continuous, i.e., $f(t) = f(t-0)$), there is a one-to-one correspondence between bounded Borel measures on $\mathbb{R}$ and functions $f \in BV_-$ as well as between Borel measures on $[a, b]$ and functions of bounded variation on $[a, b]$ with $f(a) = 0$.

Chapter 3
Fourier Series

This chapter is not a mini-course on Fourier series. If it were it would aim to cover all the basics of this topic, that is, all the basic results and their proofs, or at least give the links to them. We can say that since it is not, this chapter is a concise introduction only to certain properties of Fourier series. We are not going to cover or even mention all the features and peculiarities of this machinery. We are traveling along the path that leads to the appearance of the Fourier transform not as a non-periodic extension of the Fourier series but as an indispensable tool for clarifying or solving certain periodic problems. Of course, just a scheme is outlined, some topics are included either because they are really related, or just because they are dear to the author's taste or memory. In other words, here the reader will find either the facts and results the author dealt with or those that impressed the author in the very beginning of his mathematical biography or later on. A strong impression is sometimes important enough for including a selected topic. And it is not a novelty that dealing with Fourier series stimulates the discovery of results in more or even less related topics: this happened to Dirichlet, Riemann, and Weierstrass, to mention some.

One cannot say that the oscillating infinite sums we are going to discuss are erroneously named after Fourier; however, certain aspects were known to Daniel Bernoulli and Euler much earlier. There are many sources of this topic. The older and classical books by Zygmund [57] and Bary [4] still retain their value and importance. Among recent books, Montgomery's "Early Fourier Analysis" [39] really stands out.

3.1 Definition and Basic Properties

There are various ways to come to the idea of Fourier series. The author obtained his initial knowledge on this topic verbally from the lectures of his advisor R.M. Trigub

© The Author(s), under exclusive license to Springer Nature Switzerland AG 2021
E. Liflyand, *Harmonic Analysis on the Real Line*, Pathways in Mathematics,
https://doi.org/10.1007/978-3-030-81892-0_3

and from reading the celebrated course by Kolmogorov and Fomin [29]. There are various approaches to the matter, mainly depending on what class of functions is considered. The L^2 theory suggests the following. Let the functions be considered in the $L^2[-\pi, \pi]$ space, with respect to the usual Lebesgue measure. In this space, the functions

$$1, \cos kt, \sin kt, \qquad k = 1, 2, \ldots,$$

form a complete orthogonal system, which is called *trigonometric*. The completeness of the system (that is, the property of the system that it cannot be replenished, or the only function orthogonal to all the elements of the system is zero function) follows from the possibility to approximate any continuous periodic function by trigonometric polynomials, that is, finite linear combinations of the elements of the system in question. It is easy to see that the system is not normed. To make it orthonormal, the constant should be $\frac{1}{\sqrt{2\pi}}$, and the rest of the elements should be divided by $\frac{1}{\sqrt{\pi}}$. We can represent the considered function f by means of its Fourier series

$$\frac{a_0}{2} + \sum_{k=1}^{\infty} (a_k \cos kt + b_k \sin kt), \tag{3.1}$$

where

$$a_k = \frac{1}{\pi} \int_{-\pi}^{\pi} f(t) \cos kt \, dt, \qquad k = 0, 1, 2, \ldots,$$

and

$$b_k = \frac{1}{\pi} \int_{-\pi}^{\pi} f(t) \sin kt \, dt, \qquad k = 1, 2, \ldots,$$

are called the Fourier coefficients. In fact, the notion of the Fourier or trigonometric series suggest that the functions should be considered as 2π-periodic. Of course, the orthonormal system can be rescaled in such a way that the period can be different. For various reasons, we prefer to take it 2π. It follows from the general theory that this series converges to f in the L^2 norm. It is important that in this way the Fourier coefficients are well-defined, since an $L^2[-\pi, \pi]$ function is also Lebesgue integrable. The latter is a simple consequence of the Cauchy-Bunyakovskii-Schwarz inequality. Let us mention that Lebesgue's differentiation Theorem 2.10 holds true in the periodic case as well. As usual, by convergence we mean the convergence of the partial sums

$$S_N(t) := S_N(f; t) = \frac{a_0}{2} + \sum_{k=1}^{N} (a_k \cos kt + b_k \sin kt).$$

It follows from orthogonality that the mean square deviation of S_N from f is

$$\| f - S_N \|_2^2 = \| f \|_2^2 - \pi \left(\frac{|a_0|^2}{2} + \sum_{k=1}^{N} (|a_k|^2 + |b_k|^2) \right).$$

Among all trigonometric polynomials, that is, trigonometric polynomials of the form

$$T_N(t) = \frac{a_0}{2} + \sum_{k=1}^{N} (\alpha_k \cos kt + \beta_k \sin kt),$$

just the partial sum S_N gives the best L^2 approximation to f. The general theory suggests the Bessel inequality of the form

$$\frac{|a_0|^2}{2} + \sum_{k=1}^{\infty} (|a_k|^2 + |b_k|^2) \leq \frac{1}{\pi} \int_{-\pi}^{\pi} |f(t)|^2 \, dt.$$

However, since the trigonometric system is complete, any L^2 function enjoys the Parseval identity

$$\frac{|a_0|^2}{2} + \sum_{k=1}^{\infty} (|a_k|^2 + |b_k|^2) = \frac{1}{\pi} \int_{-\pi}^{\pi} |f(t)|^2 \, dt.$$

The Fischer–Riesz theorem supplements this statement with the converse assertion that if the numbers $a_k, b_k, k = 0, 1, 2, \ldots$, are such that the number series

$$\sum_{k=1}^{\infty} (|a_k|^2 + |b_k|^2) < \infty,$$

then the series

$$\frac{a_0}{2} + \sum_{k=1}^{\infty} (a_k \cos kt + b_k \sin kt)$$

not only converges in L^2 but also its sum as a function has the numbers a_k, b_k as the Fourier coefficients of this function.

Square summability of the Fourier coefficients is essential. A notion of Carleman singularity is of crucial importance: there is a continuous periodic function for which the sequence of its Fourier coefficients is not p-summable for any $p < 2$; such a construction can be found, for instance, in [57, Chapter V, §5, (4.11)]. This is the property not only of the trigonometric system but of any complete orthonormal system, see [176].

There are other trigonometric systems related to the one considered above. First of all, the exponential system

$$e^{ikt}, \qquad k = 0, \pm 1, \pm 2, \dots$$

Because of the Euler identity $e^{ikt} = \cos kt + i \sin kt$, each of the two systems can be rewritten in the form of the other. The corresponding Fourier series can be rewritten in the complex form

$$\sum_{k=-\infty}^{\infty} c_k e^{ikt}, \tag{3.2}$$

where

$$c_k = \frac{1}{2\pi} \int_{-\pi}^{\pi} f(t) e^{-ikt} \, dt.$$

This form is more modern and usable, while (3.1) is somewhat out of fashion, though the two are just equivalent by denoting $c_k = \alpha_k + i\beta_k$. The corresponding calculations are trivial. This and the Euler identity allow one to rewrite the same Fourier series in either of the two forms. The completeness and other properties do not depend on the series notation. The N-th partial sum in the latter case is

$$S_N = S_N(f; t) = \sum_{k=-N}^{N} c_k e^{ikt}. \tag{3.3}$$

Finally, the systems

$$1, \cos t, \cos 2t, \dots$$

and

$$\sin t, \sin 2t, \dots$$

can be considered separately but on the interval $[0, \pi]$. Their completeness is related to the even extension of f from $[0, \pi]$ to $[-\pi, \pi]$, and to the odd one, respectively. In fact, since every function can be represented as the sum of its even part and odd part by

$$f(t) = \frac{f(t) + f(-t)}{2} + \frac{f(t) - f(-t)}{2} := f_e(t) + f_o(t), \tag{3.4}$$

in the Fourier expansion the summands with cosine waves correspond to the even part, while those with the sine waves correspond to the odd part.

3.2 Convergence

What made the Fourier series a solid topic rather than just a convenient tool are convergence and divergence properties of these series for functions from different classes. For the basics, we follow the way and order in which it is given in [18], since the amount of material and the form of the presentation fit the scope of our book. Given the Fourier series of f in the form (3.2), its N-th partial sum (3.3) can be written as

$$S_N(x) = \frac{1}{2\pi} \int_{-\pi}^{\pi} f(t) \sum_{k=-N}^{N} e^{ik(x-t)} \, dt.$$

Denoting by

$$D_N(t) = \sum_{k=-N}^{N} e^{ikt} \tag{3.5}$$

the *Dirichlet kernel*, we see that the partial sum is the convolution of the function and the Dirichlet kernel:

$$S_N(x) = \frac{1}{2\pi} \int_{-\pi}^{\pi} f(t) D_N(x-t) \, dt. \tag{3.6}$$

In fact, the name Dirichlet kernel is more frequently given to

$$D_N^c(t) = \frac{1}{2} + \sum_{k=1}^{N} \cos kt = \frac{\sin(N+\frac{1}{2})t}{2\sin\frac{t}{2}}. \tag{3.7}$$

Also, the conjugate Dirichlet kernel

$$D_N^s(t) = \sum_{k=1}^{N} \sin kt = \frac{\cos\frac{t}{2} - \cos(N+\frac{1}{2})t}{2\sin\frac{t}{2}} \tag{3.8}$$

is used systematically. It is clear from the definition that D_N is an even 2π-periodic function and that

$$\int_{0}^{\pi} D_N(t) \, dt = \pi.$$

Since

$$(e^{i\frac{t}{2}} - e^{-i\frac{t}{2}})D_N(t) = e^{it(N+\frac{1}{2})} - e^{-it(N+\frac{1}{2})},$$

we have

$$D_N(t) = \frac{\sin(N + \frac{1}{2})t}{\sin \frac{t}{2}}, \quad t \neq 2k\pi. \tag{3.9}$$

Now, changing the variables $u = t - x$, using the periodicity of both f and D_N, and the fact that D_N is an even function, we obtain

$$S_N(x) = \frac{1}{2\pi} \int_0^\pi [f(x+t) + f(x-t)]\, D_N(t)\, dt. \tag{3.10}$$

Therefore, the sequence $\{S_N(x)\}$ converges to the number $v(x)$ if and only if

$$0 = \lim_{N \to \infty} [S_N(x) - v(x)]$$

$$= \lim_{N \to \infty} \frac{1}{2\pi} \int_0^\pi [f(x+t) + f(x-t) - 2v(x)]\, D_N(t)\, dt.$$

Combining all the formulas and denoting

$$g(x,t) = f(x+t) + f(x-t) - 2v(x), \tag{3.11}$$

we have the following

Theorem 3.12 *If f is a 2π-periodic function, integrable on the period, then the sequence $\{S_N(x)\}$ converges to the number $v(x)$ if and only if*

$$\lim_{N \to \infty} \int_0^\pi g(x,t) D_N(t)\, dt = \lim_{N \to \infty} \int_0^\pi \frac{g(x,t)}{\sin \frac{t}{2}} \sin\left(N + \frac{1}{2}\right)t\, dt = 0. \tag{3.13}$$

Details of pointwise convergence that will be discussed below are strongly based on the following Riemann–Lebesgue lemma. This fact, in a more general setting will be given in the next chapter as Lemma 4.3. However, for convenience, we present this fact here in one of the possible forms:

If $f \in L^1[a,b]$, $-\infty \leq a < b \leq \infty$,   then $\displaystyle\lim_{|x| \to \infty} \int_a^b f(t)e^{ixt}\,dt = 0.$

What immediately follows from the Riemann–Lebesgue lemma is that the Fourier coefficients tend to zero. However, this tendency can be arbitrarily slow; we discuss this issue in more detail in the chapter about the Fourier transform. When certain additional assumptions are posed on the function, the Fourier coefficients may decay faster. For instance, we shall see this while discussing conditions that guarantee the absolute convergence of Fourier series. A different example is as follows. Many years ago, the author was impressed by Taibleson's paper [190] where a simple and known result was proved in an extremely brief way. This reading also resulted in a long lasting desire of the author to publish a paper in as brief a way as possible; now hopefully this has almost been accomplished in [156]. In order to present Taibleson's result, we deal with the Fourier series in the complex form (3.2).

Proposition 3.14 *If f is of bounded variation on $\mathbb{T}$, then $c_k = O\left(\frac{1}{k}\right)$ as $|k| \to \infty$.*

Proof Supposing that $k \neq 0$, for fixed k, we set $d_j = \frac{2\pi j}{|k|}$, $j = 0, 1, \ldots, |k|$. Let g be a step function equal to $f(d_j)$ on (d_{j-1}, d_j), $j = 1, \ldots, |k|$. Since

$$\int_{\frac{2\pi j}{|k|}}^{\frac{2\pi(j+1)}{|k|}} e^{-ikt}\, dt = 0,$$

and this is the main trick in the proof, we have

$$\frac{1}{2\pi} \int_0^{2\pi} g(t) e^{-ikt}\, dt = 0.$$

Rewriting

$$c_k = \frac{1}{2\pi} \int_0^{2\pi} f(t) e^{-ikt}\, dt,$$

we obtain

$$|c_k| = \left| \int_0^{2\pi} [f(t) - g(t)] e^{-ikt}\, dt \right|$$

$$\leq \sum_{j=1}^{|k|} \int_{d_{j-1}}^{d_j} |f(t) - f(d_j)|\, dt$$

$$\leq \sum_{j=1}^{|k|} V_{(d_{j-1}, d_j)} f \, \frac{2\pi}{|k|} = \frac{2\pi}{|k|} V_{\mathbb{T}} f,$$

as required. $\qquad\qquad\qquad\square$

For the convergence, the Riemann–Lebesgue lemma gives the following modification of (3.13): the equivalent necessary and sufficient condition in Theorem 3.12 is that, for some $\delta \in (0, \pi)$,

$$\lim_{N \to \infty} \int_0^\delta \frac{g(x, t)}{\sin \frac{t}{2}} \sin\left(N + \frac{1}{2}\right) t\, dt = 0. \qquad (3.15)$$

The Riemann–Lebesgue lemma just "eats" the integral over (δ, π). Further, since $\sin \frac{t}{2}$ is equivalent to t, (3.15) can be somewhat simplified as

$$\lim_{N \to \infty} \int_0^\delta \frac{g(x, t)}{t} \sin\left(N + \frac{1}{2}\right) t \, dt = 0. \tag{3.16}$$

The Riemann–Lebesgue lemma leads to a sufficient condition for the convergence of the Fourier series at x. This condition, the integrability of $\dfrac{g(x, t)}{t}$ on $[0, \delta]$, for some $\delta \in (0, \pi)$, is known as Dini's condition. Another useful result on pointwise convergence of Fourier series is the Jordan test (a particular case is due to Dirichlet).

Theorem 3.17 *If f is an integrable 2π-periodic function, which is of bounded variation on $[a, b] \subset [-\pi, \pi]$, then the Fourier series of f converges at x to*

$$v(x) = \frac{f(x+) + f(x-)}{2}$$

for each $x \in (a, b)$.

Proof Let $0 < \delta < \min\{x - a, b - x, \pi\}$. Then $g(x, t)$ is of bounded variation on $[0, \delta]$ as a function of t. Therefore, for x fixed, g is the difference of two nondecreasing functions g_1 and g_2 on $[0, \delta]$:

$$g(x, t) = g_1(t) - g_2(t), \qquad t \in [0, \delta].$$

Since

$$0 = g(x, 0+) = g_1(0+) - g_2(0+),$$

if we let

$$h_j(t) = g_j(t) - g_j(0+), \qquad j = 1, 2,$$

then $h_j(0+) = 0$, each of h_j is nondecreasing on $[0, \delta]$ and nonnegative on $(0, \delta]$, and

$$g(x, t) = h_1(t) - h_2(t), \qquad t \in [0, \delta].$$

By (3.16), we have that $S_N(x)$ converge to

$$v(x) = \frac{f(x+) + f(x-)}{2}$$

if

$$\lim_{N \to \infty} \int_0^{\delta} \frac{h_j(t)}{t} \sin\left(N + \frac{1}{2}\right)t \, dt = 0, \qquad j = 1, 2.$$

For any small $\varepsilon > 0$, since $h_j(0+) = 0$, there exists an $\eta \in (0, \delta)$ such that $h_j(t) < \varepsilon$ for $t \in (0, \eta)$. By the Second Mean Value Theorem, for some $c \in [0, \eta]$ and for all N, we have

$$\left| \int_0^{\eta} \frac{h_j(t)}{t} \sin(N + \frac{1}{2})t \, dt \right| = \left| h_j(\eta-) \int_c^{\eta} \frac{\sin(N + \frac{1}{2})t}{t} \, dt \right|$$

$$< \varepsilon \left| \int_{(N+\frac{1}{2})c}^{(N+\frac{1}{2})\eta} \frac{\sin t}{t} \, dt \right| \leq 4\varepsilon.$$

Since by the Riemann–Lebesgue lemma,

$$\left| \int_{\eta}^{\delta} \frac{h_j(t)}{t} \sin\left(N + \frac{1}{2}\right)t \, dt \right| < \varepsilon,$$

for N large enough, we get

$$\left| \int_0^{\delta} \frac{h_j(t)}{t} \sin\left(N + \frac{1}{2}\right)t \, dt \right| < 5\varepsilon,$$

for N large enough, as desired.   □

Old results of Kolmogorov show that for convergence at a point the function must be good enough, or, more precisely, not bad enough. In [132], he gave an example of an integrable function whose Fourier series diverges almost everywhere; later on, in [133], he refined this example by finding an integrable function whose Fourier series diverges *everywhere*!

At the dawn of the history of harmonic analysis, it had been thought that continuous functions surely have convergent Fourier series. It came as a surprise when in 1873, du Bois Reimond gave an example of a continuous function whose Fourier series diverges at a point. The following construction was one of my first impressions on the depth and attraction of Fourier Analysis when I was a student at Trigub's lectures. We outline the proof as it is given in [44, 1.6.1]. Let

$$f(t) = \sum_{k=1}^{\infty} e^{iN_k t} \sum_{j=1}^{2^{k^3}} \frac{\sin jt}{j},$$

where one can take $N_1 = 1$ and for $k = 1, 2, \ldots,$

$$N_{k+1} = \sum_{j=0}^{k-1} (2^{j^3} + 2^{(j+1)^3} + 1).$$

By this, one can see that the terms of the inner sums contain different frequencies. Also, the m-th Fourier coefficient of f is either

$$\frac{1}{(m - N_k)k^2}$$

or zero. Then convergence of the subsequence of partial sums $S_{N_k}(0)$ depends on the behavior of

$$\frac{1}{k^2} \sum_{j=1}^{2^{k^3}} \frac{1}{j},$$

which is equivalent to k. Therefore, this subsequence diverges.

Of course, divergence at zero is taken for convenience, it can be any point. In fact, much later Kahane and Katznelson in [128] showed that any preassigned set of Lebesgue measure zero can be the set of divergence of the Fourier series of a continuous function.

In his thesis of 1915 (published only in 1951 as [36]), Luzin conjectured, as opposed to the L^1 functions, that the Fourier series of a continuous function (wider—of a square integrable function) converges almost everywhere. This conjecture remained open till 1966 when Carleson gave his remarkable and, in fact, very complicated proof in [84]. Shortly thereafter, Hunt [123] extended this result to the L^p functions, $1 < p < \infty$. In [98], Ch. Fefferman gave a proof based on different ideas. The proof by Lacey and Thiele in [138] was a sort of refinement of Fefferman's ideas and was presented in an extremely short form. A good source for a detailed consideration of their arguments is [16, Chapter 11]. One more proof, related to Fefferman's, was recently given in [175], with the ideas and techniques of bounded r-variation widely exploited. There are further general works where it is claimed that the solution of Luzin's problem comes as a partial case. The most comprehensive among those is [201]. All in all, a heuristic principle can be proclaimed: *The smoother the function, the more rapidly its Fourier series converges, and vice-versa.* One can see that this intriguing story has lasted for more than 100 years and probably will go on and on.

3.3 Absolute Convergence

There are various types of convergence of Fourier series. Each suggest a variety of interesting problems as well as relations between them. For example, §44 in Chapter I of [4] is devoted to an example of a function with the Fourier series which converges everywhere but not uniformly. While the above example of a continuous function with the Fourier series divergent at a point was my first strong impression of the topic, the results on the absolute convergence of Fourier series, especially those due to Bernstein and Zygmund, were my first love in Fourier Analysis.

In general, absolute convergence of a series means the convergence of the series of the absolute values of the terms. For a series of the form (3.2) this is equivalent to

$$\sum_{k=-\infty}^{\infty} |c_k| < \infty.$$

If one deals with a series of the form (3.1), there appears certain problems related to the argument but we shall set them aside and understand the absolute convergence in this case as

$$\sum_{k=1}^{\infty} (|a_k| + |b_k|) < \infty.$$

At first glance, the functions with absolutely convergent Fourier series are too well-defined to be interesting (for instance, such a Fourier series also converges uniformly), however there are many delicate and hard problems in this topic. This topic goes back a long time if we date it from Bernstein's theorem (see the second part of Theorem 3.18 below), which was proved in 1905. However, not only is the topic active today but it never was considered as old-fashioned. Kahane's book [25], though never translated into English, is still a source of many interesting ideas.

Very often functions with absolutely convergent Fourier series are treated as a Banach space A endowed with the norm

$$\|f\|_A = \sum_{k \in \mathbb{Z}} |c_k|,$$

for $f \in A$, whose Fourier series are of the form (3.2). In addition, such functions form a Banach algebra, with $\|fg\|_A \leq \|f\|_A \|g\|_A$. In particular, one of the most impressive outcomes of the theory of Banach algebras, when invented about 80 years ago, was an immediate proof of Wiener's famous result on the absolute convergence of the Fourier series of $\frac{1}{f}$ for never vanishing $f \in A$.

We now present the mentioned Bernstein's theorem, or, more precisely, one of its numerous versions.

Theorem 3.18 *Let f be a 2π-periodic function with the modulus of continuity $\omega(f; h)$. If the series*

$$\sum_{k=1}^{\infty} \frac{\omega(f; \frac{1}{k})}{\sqrt{k}} \tag{3.19}$$

converges, then the Fourier series of f converges absolutely.

In particular, if f is of Lip α *smoothness, with $\alpha > \frac{1}{2}$, then the Fourier series of f converges absolutely.*

Zygmund's theorem differs from this one in the sense that here the function is supposed to be of bounded variation.

Theorem 3.20 *Let f be a 2π-periodic function of bounded variation, with the modulus of continuity $\omega(f; h)$. If the series*

$$\sum_{k=1}^{\infty} \frac{\sqrt{\omega(f; \frac{1}{k})}}{k} \tag{3.21}$$

converges, then the Fourier series of f converges absolutely.

In particular, if f is of Lip α *smoothness, with $\alpha > 0$, then the Fourier series of f converges absolutely.*

In both theorems the second parts are readily derived from the first parts. The latter ones, in turn, follow from the following theorem of Szász.

Theorem 3.22 *Let f be a 2π-periodic function with the modulus of continuity $\omega(f; h)_2$. If the series*

$$\sum_{m=1}^{\infty} \frac{\omega(f; \frac{1}{m})_2}{\sqrt{m}} \tag{3.23}$$

converges, then the Fourier series of f converges absolutely.

It is claimed in [4] that the following proof goes along the lines that are due to Stechkin.

Proof Starting from (3.2), we have

$$f(t+h) - f(t-h) \sim \sum_{k=-\infty}^{\infty} c_k e^{ikt}[e^{ikh} - e^{-ikh}] = 2i \sum_{k=-\infty}^{\infty} c_k e^{ikt} \sin kh.$$

Parseval's identity yields

$$\frac{1}{2\pi} \int_{-\pi}^{\pi} |f(t+h) - f(t-h)|^2 \, dt = 4 \sum_{k=-\infty}^{\infty} |c_k|^2 \sin^2 kh. \tag{3.24}$$

Let us integrate in h over $[0, \frac{\pi}{m}]$. We get

$$\int_0^{\frac{\pi}{m}} \frac{1}{2\pi} \int_{-\pi}^{\pi} |f(t+h) - f(t-h)|^2 \, dt \, dh = 4 \sum_{k=-\infty}^{\infty} |c_k|^2 \int_0^{\frac{\pi}{m}} \sin^2 kh \, dh.$$

For the given k, we choose an integer l such that $l \le \frac{k}{m} < l+1$. By substitution $kh = s$, the last integral is estimated as

$$\frac{1}{k} \int_0^{\frac{k\pi}{m}} \sin^2 s \, ds \ge \frac{1}{m(l+1)} \int_0^{l\pi} \sin^2 s \, ds$$

$$= \frac{l}{m(l+1)} \int_0^{\pi} \sin^2 s \, ds = \frac{l\pi}{2m(l+1)}.$$

If $|k| \ge m$, that is $l \ge 1$, then $\frac{l}{l+1} \ge \frac{1}{2}$. Finally,

$$4 \int_0^{\frac{\pi}{m}} \sin^2 kh \, dh \ge \frac{\pi}{m}.$$

Combining the last calculations, we obtain, for $0 < h \le \frac{\pi}{m}$,

$$\sum_{|k| \ge m} |c_k|^2 \le \frac{m}{\pi} \int_0^{\frac{\pi}{m}} \frac{1}{2\pi} \int_{-\pi}^{\pi} |f(t+h) - f(t-h)|^2 \, dt \, dh$$

$$\le \omega\left(f; \frac{\pi}{m}\right)_2^2 \le C\omega\left(f; \frac{1}{m}\right)_2^2.$$

From this it follows that

$$\sum_{|k| \ge 1} |c_k| \le C \sum_{|k| \ge 1} \sum_{m \le |k|} \frac{|c_k|}{|k|} \le C \sum_{m=1}^{\infty} \sum_{|k| \ge m} \frac{|c_k|}{|k|}.$$

Now, the Cauchy-Bunyakovskii-Schwarz inequality yields

$$\sum_{|k|\geq 1} |c_k| \leq C \sum_{m=1}^{\infty} \sqrt{\sum_{|k|\geq m} \frac{1}{k^2} \sum_{|k|\geq m} |c_k|^2}$$

$$\leq C \sum_{m=1}^{\infty} \frac{\omega(f;\frac{1}{m})2}{\sqrt{m}},$$

which completes the proof. $\square$

In fact, this proof can be shortened if after (3.24), one continues with $h = \frac{\pi}{2^m}$ without integration.

There is one more principal difference between these two types of results. While Bernstein type results had a number of sharpness constructions (see, e.g., [25]), the sharpness of Zygmund type results was proved only once by means of very sophisticated tools (see [8, 76]). More precisely, in the cited works, assuming that the series in (3.21) diverges for a given modulus of continuity type function, Bochkarev constructed a function of bounded variation with the same modulus of continuity and with the Fourier series not absolutely convergent.

In fact, an idea was suggested some time ago that most of the conditions for the absolute convergence of Fourier series stem from the assumption that the function satisfies two smoothness conditions, say, of degree α and β, with $\alpha + \beta > 1$, where 1 is the dimension of the space. When we have one condition, it can be interpreted as the same two conditions sticking together. This is the case for Bernstein type results, we just have, for $\alpha > \frac{1}{2}$, that $\alpha + \alpha > 1$. As for Zygmund type conditions, we have α Lipschitz smoothness, $\alpha > 0$, and boundedness of variation, which can be treated as Lip 1 smoothness in the L^1 metrics, whereas $\alpha + 1 > 1$ follows. This heuristic idea is developed for a variety of pairs of spaces in [160]; the results are derived for the Fourier transform in the n-dimensional case. The former is not that important, usually such results for the Fourier transform and Fourier series are similar. The sufficient relation in the multivariate case is, roughly speaking, $\alpha + \beta > n$. It is worth noting that in such a situation *interpolation* is one of the main tools.

3.4   Lebesgue Constants

The above discussion of the fact that the Fourier series of a continuous function need not be convergent at a specific point can be solved on the functional-analytic level rather than by means of counterexamples. By the Banach-Steinhaus theorem, it suffices to show that the sequence of the norms of the partial sums operators (taking C into C or L^1 into L^1, which is the same) is not bounded. Such norms are called the *Lebesgue constants*. Let us derive the classical asymptotic formula for them.

Starting with (3.10), we outline how one can obtain a formula for the norms L_N of S_N as an operator taking C into C. Considering the value of the operator on the function sign $D_N(t)$, we immediately arrive at the expression

$$L_N = \frac{1}{\pi} \int_0^\pi |D_N(t)| \, dt. \tag{3.25}$$

The formula is true but is not yet proven, since sign $D_N(t)$ is not a continuous function. However, it can be corrected on a small set to be continuous. Getting in this way a sequence of continuous functions, we can approach (3.25). We will now prove that

$$L_N = \frac{4}{\pi^2} \ln N + O(1). \tag{3.26}$$

In fact, this particular formula comes to mind to almost every mathematician when hearing or reading the words "Lebesgue constants".

Since

$$D_N(t) = 2 \frac{\sin Nt}{t} + O(1),$$

we have

$$\frac{2}{\pi} \int_0^\pi \frac{|\sin Nt|}{t} \, dt + O(1).$$

The proof now reduces to obtaining the corresponding asymptotic relation for the value

$$I_N = \int_0^\pi \frac{|\sin Nt|}{t} \, dt = \int_0^{N\pi} \frac{|\sin t|}{t} \, dt.$$

Denoting analogously

$$I_k = \int_0^\pi \frac{|\sin kt|}{t} \, dt = \int_0^{k\pi} \frac{|\sin t|}{t} \, dt,$$

we have

$$I_{k+1} - I_k = \int_{k\pi}^{(k+1)\pi} \frac{|\sin t|}{t} \, dt = \int_0^\pi \frac{\sin t}{t + k\pi} \, dt.$$

Since for $0 \leq t \leq \pi$, we have

$$\frac{1}{\pi(k+1)} \leq \frac{1}{t + k\pi} \leq \frac{1}{k\pi},$$

the relation

$$\int_0^\pi \sin t \, dt = 2$$

yields

$$\frac{2}{\pi(k+1)} \le I_{k+1} - I_k \le \frac{2}{k\pi}.$$

Summing these inequalities for $k = 1, 2, \ldots, N - 1$, we get

$$\frac{2}{\pi} \sum_{k=1}^{N-1} \frac{1}{k+1} \le \sum_{k=1}^{N-1} (I_{k+1} - I_k) \le \frac{2}{\pi} \sum_{k-1}^{N-1} \frac{1}{k},$$

or, equivalently,

$$I_1 + \sum_{k=2}^{N} \frac{1}{k} \le I_N \le I_1 + \sum_{k=1}^{N-1} \frac{1}{k}.$$

Since both sums are equivalent to $\ln N$, we arrive at (3.26).

We shall see shortly (in the following section) how to get this asymptotic formula as a particular case of a more general result.

Meanwhile, an immediate application follows.

Proposition 3.27 *Let f be a bounded periodic function, say, $|f(t)| \le M$, for all t. Then, for N large and each x, we have*

$$|S_N(f; x)| \le \frac{2M}{\pi^2} \ln N.$$

Proof By (3.6), there holds

$$|S_N(f; x)| \le \frac{1}{2\pi} \int_{-\pi}^{\pi} |f(t)| \left| D_N(x - t) \right| dt \le \frac{M}{2} \frac{1}{\pi} \int_{-\pi}^{\pi} \left| D_N(t) \right| dt.$$

By (3.26), we have

$$|S_N(f; x)| \le \frac{M}{2} \frac{4}{\pi^2} \ln N,$$

for N large enough, as required.   □

3.5 Summability

It has been known for a long time that divergent number series can still be studied by means of summability methods; the book [21] is seminal for this approach. For the Fourier series, the result of Féjer is fundamental in the study of summability methods. Though the Fourier series of an integrable function can be drastically divergent, Féjer's discovery that taking the arithmetic means of the partial sums of (divergent) Fourier series always leads to convergence results remains in force for even unappropriate, in a sense, objects. More precisely, we consider

$$\sigma_N(t) = \sigma_N(f; t) = \frac{1}{N} \sum_{k=0}^{N-1} S_k(x). \tag{3.28}$$

These means can be represented like (3.6) as a convolution but with a different kernel. This Féjer kernel $K_N(t)$ can easily be calculated in the same way as the Dirichlet kernel:

$$K_N(t) = \frac{1}{2N} \left(\frac{\sin \frac{Nt}{2}}{\sin \frac{t}{2}} \right)^2. \tag{3.29}$$

The positivity of this kernel plays a crucial role in many problems.

The Féjer means $\sigma_N(f; t)$ converge to the value of the function at a continuity point (uniformly if the function is everywhere continuous), to

$$\frac{f(t+) + f(t-)}{2}$$

if t is a point of discontinuity of the first kind, and almost everywhere for an integrable function (in fact, at each Lebesgue point). It is usually said in this case that the series is Cesàro or $(C, 1)$ summable. Instead of proving this directly we shall show a more general approach to this and many other problems of summability of Fourier series. One can easily see that the Féjer means enjoy a different representation:

$$\sigma_N(t) = \sigma_N(f; t) = \sum_{k=-N}^{N} \left(1 - \frac{|k|}{N} \right) c_k e^{ikt},$$

and $1 - \frac{|k|}{N}$ are the values $\lambda(\frac{|k|}{N})$ for the function $\lambda(t) = (1 - |t|)_+$ (the latter symbol means that the function is zero when the value in the parentheses is negative). Many other functions λ are considered in this way, and they are called *multipliers*. The corresponding means

$$\Lambda_N(t) = \Lambda_N(f; t) = \sum_{k=-\infty}^{\infty} \lambda\left(\frac{k}{N}\right) c_k e^{ikt} \tag{3.30}$$

are meaningful under appropriate convergence conditions. By orthonormality, we can rewrite (3.30), at least formally, as

$$\Lambda_N(t) = \Lambda_N(f; t) = \sum_{k=-\infty}^{\infty} \lambda\left(\frac{k}{N}\right) e^{ikt} \frac{1}{2\pi} \int_{-\pi}^{\pi} f(u) e^{-iku} \, du$$

$$= \frac{1}{2\pi} \int_{-\pi}^{\pi} f(t \mid u) \sum_{k=-\infty}^{\infty} \lambda\left(\frac{k}{N}\right) e^{-iku} \, du.$$

Substituting $Nu = s$, we have

$$\Lambda_N(t) = \frac{1}{2\pi} \int_{-N\pi}^{N\pi} f(t + \frac{s}{N}) \left[\sum_{k=-\infty}^{\infty} \lambda\left(\frac{k}{N}\right) e^{-i\frac{k}{N}s} \frac{1}{N} \right] ds.$$

One can observe that the expression in the brackets,

$$\sum_{k=-\infty}^{\infty} \lambda\left(\frac{k}{N}\right) e^{-i\frac{k}{N}s} \frac{1}{N},$$

is similar in appearance to the integral sum of the following integral:

$$\int_{\mathbb{R}} \lambda(y) e^{-isy} \, dy.$$

This is neither more nor less than the Fourier transform of λ. We will devote the next chapter to this important operator of harmonic analysis. Meanwhile we will introduce only the notation

$$\widehat{\lambda}(s) := \int_{\mathbb{R}} \lambda(y) e^{-isy} \, dy,$$

without discussing whether and when this integral converges and what its properties are. In the same informal manner, we have

$$\Lambda_N(t) = \frac{1}{2\pi} \int_{-N\pi}^{N\pi} f(t + \frac{s}{N}) \widehat{\lambda}(s) \, ds + \text{possible remainder terms}.$$

The precise form of the latter relation depends, of course, on the class on which we consider the problem and on the properties of the multiplier λ. We mention only that considering the norms Λ_N of the sequence of operators taking C into C, we have for a wide class of λ-s

$$L_N = \frac{1}{2\pi} \int_{-\pi}^{\pi} \left| \sum_{k=-\infty}^{\infty} \lambda\left(\frac{k}{N}\right) e^{iku} \right| du. \qquad (3.31)$$

These numbers are also called (generalized) Lebesgue constants; most of the known results on them (in the multivariate case) are surveyed in [144]. Under additional natural assumptions on λ, we have (see [69])

$$L_N = \frac{1}{2\pi} \int_{\mathbb{R}} |\widehat{\lambda}(s)|\, ds + \frac{\theta}{2\pi} \int_{|s| \geq N\pi} |\widehat{\lambda}(s)|\, ds, \qquad -2 \leq \theta \leq 0, \quad (3.32)$$

where θ, generally speaking, depends on N, and

$$\sup_N L_N = \frac{1}{2\pi} \int_{\mathbb{R}} |\widehat{\lambda}(s)|\, ds. \qquad (3.33)$$

The main assumption for all such results is the Lebesgue integrability of $\widehat{\lambda}$. This is certainly true for the Féjer (Cesáro) means. Many of such results can be found in [55] and [161]; they will also be studied in the following chapter.

However, Λ_N can often be expressed in terms of $\widehat{\lambda}$ also in the cases where $\{\Lambda_N\}$ is an unbounded sequence (and $\widehat{\lambda}$ is not integrable). In such situations, a typical form of an estimate is

$$L_N = \frac{1}{2\pi} \int_{-N\pi}^{N\pi} |\widehat{\lambda}(s)|\, ds + \text{possible remainder terms}. \qquad (3.34)$$

Of course, such a formula makes sense if the remainder term(s) are good enough, say, uniformly bounded or behave better, with respect to N, than the leading term.

The above considered case of the Lebesgue constants for partial sums also fits within this scope. For this, λ should simply be the indicator function of an interval.

3.6   Trigonometric Series Versus Fourier Series

It is really amazing how often people do not distinguish between trigonometric series and Fourier series, using the former as a synonym of the latter. However, the difference is sometimes crucial, especially in the problem of whether the trigonometric series is the Fourier series of an integrable function.

More precisely, the problem reads as follows. Given a trigonometric series

$$\frac{a_0}{2} + \sum_{k=1}^{\infty}(a_k \cos kx + b_k \sin kx),$$

under what assumptions on its coefficients is this series the Fourier series of an integrable function? Sometimes, it is briefly said "a Fourier series" instead of "the Fourier series of an integrable function". The abuse of terminology is almost negligible in these circumstances, especially if we recall that to calculate the Fourier coefficients the function must be Lebesgue integrable. If the assumptions considered are such that the series converges to an integrable function almost everywhere, then the problem whether this series is a Fourier series is reduced to that of integrability of the sum of the given series; briefly integrability of the series. Here the abuse of notation is also non-essential, the term means integrability of the function which coincides with the sum of the series almost everywhere. In a different way, we shall say in these cases that the sequence of coefficients belongs to $\widehat{L^1}$.

Frequently, the cosine series

$$\frac{a_0}{2} + \sum_{k=1}^{\infty} a_k \cos kx \tag{3.35}$$

and the sine series

$$\sum_{k=1}^{\infty} b_k \sin kx \tag{3.36}$$

are investigated separately, since there is a difference in their behavior. Usually, obtaining results for the latter requires additional assumptions.

Of course, series of the form (3.2) have also been studied in this respect (see, e.g., [172]), but this did not add anything essential till recently.

To the best of our knowledge, a convenient description of $\widehat{L^1}$ in terms of a given sequence alone did not exist till recently. Actually, there were some characterizations, e.g., [46, 95, 181, 182], but they are too complicated to be applied to concrete problems and they involve properties of functions. Hence, certain subspaces of $\widehat{L^1}$ have been studied so that they are both as wide as possible and described in terms convenient for applications.

Let us be more precise. First of all, in view of the Riemann–Lebesgue lemma, $\widehat{L^1}$ itself is a subspace of c_0, the space of null sequences. In 1921, Szidon [186] (see also [4, Volume I]) gave an example of an even monotone null sequence which is not in $\widehat{L^1}$. This means that also the space of sequences of bounded variation

$$bv = \left\{ d = \{d_k\} : \|d\|_{bv} = \sum_{k=0}^{\infty} |\Delta d_k| < \infty \right\}$$

is not a subspace of $\widehat{L^1}$. Here $\Delta d_k = d_k - d_{k+1}$. It is well-known (see, e.g., [4] and [57]) that possessing a null sequence of bounded variation as its Fourier coefficients, the cosine series converges for every $x \neq 0 (\mod 2\pi)$, while the sine series converges everywhere. This is not difficult to see by applying one very important and popular expedient in analysis, the so-called Abel transformation:

$$\sum_{k=m}^{n} a_k b_k = \sum_{k=m}^{n-1} A_m^k (b_k - b_{k+1}) + A_m^n b_n, \qquad (3.37)$$

where $A_m^k = a_m + \ldots + a_k$. Indeed, applying (3.37) to (3.35) and (3.36), respectively, we have

$$\frac{a_0}{2} + \sum_{k=1}^{\infty} a_k \cos kx = \sum_{k=0}^{\infty} D_k^c(x) \Delta a_k \qquad (3.38)$$

and

$$\sum_{k=1}^{\infty} b_k \sin kx = \sum_{k=1}^{\infty} D_k^s(x) \Delta b_k. \qquad (3.39)$$

By the boundedness of the kernel $D_k^c(x)$ for every $x \neq 0 (\mod 2\pi)$, and of the kernel $D_k^s(x)$ for all x, and by the fact that both sequences of coefficients are bv sequences, we conclude that both series converge absolutely, which yields the desired properties.

Trigonometric series with monotone coefficients possess many nice properties, every classical source contains a chapter or section on this topic; see, e.g., [57, Chapter V]. Important additional details can be found in [37].

Going back to the origin, we mention that in 1913, Young [200] proved that if $\{a_k\}$ is a convex null sequence, that is,

$$\Delta^2 a_k = \Delta(\Delta a_k) \geq 0$$

for $k = 0, 1, 2, \ldots$, then the cosine series is the Fourier series of an integrable function. That work and that year are traditionally considered to be the starting point for the topic in question. Of course, the class of convex sequences is a subspace of bv. Inspired by Szidon's observation, only subspaces of bv continued to be studied, that is, the mentioned problem of integrability of trigonometric series to be solved, a very important part but only a part of the general problem. In 1923, Kolmogorov [132] extended Young's result to the class of quasi-convex sequences $\{a_k\}$, namely those satisfying

$$\sum_{k=0}^{\infty} (k+1) |\Delta^2 a_k| < \infty.$$

In 1934, Pfleger [179] proved that similar to the fact that every real sequence of bounded variation is a difference of two monotone sequences, every real quasi-convex sequence is a difference of two convex sequences.

The first period of investigation ended, in a sense, in 1956 when Boas generalized all previous results in [74]. In the following years, extensions of Boas' results were studied.

Let us give a later and more contemporary list of sequence spaces all of which are subspaces of $\widehat{L^1}$ and ensure the integrability of the corresponding trigonometric series. This list does not pretend to be comprehensive. Though most of the strongest known conditions are in this list, the selection is partly a matter of taste.

1. The so-called Boas-Telyakovskii condition (see, e.g., [191]):

$$s_d = \sum_{n=2}^{\infty} \left| \sum_{k=1}^{[\frac{n}{2}]} \frac{\Delta d_{n-k} - \Delta d_{n+k}}{k} \right|, \tag{3.40}$$

where $[\alpha]$ means the integer part of the number α. This is a generalization of Boas' result in the way that in [74] the sign of absolute value in the representation for s_d was inside the second sum.

2. Fomin's condition [105] (see also [110]):

$$\|d\|_{o_p} = \sum_{n=0}^{\infty} 2^{\frac{n}{p'}} \left\{ \sum_{k=2^n}^{2^{n+1}-1} |\Delta d_k|^p \right\}^{\frac{1}{p}} < \infty, \qquad 1 < p < \infty, \quad \frac{1}{p} + \frac{1}{p'} = 1. \tag{3.41}$$

3. The Szidon-Telyakovskii condition [192]:
   There exists a sequence $\{A_k\}$ such that

$$A_k \downarrow 0 \ (k \to \infty), \qquad \sum_{k=0}^{\infty} A_k < \infty \qquad \text{and} \qquad |\Delta d_k| < A_k. \tag{3.42}$$

   Here the space in (3) is a subspace of any sequence space in (2). In turn, all of them are subspaces of the space in (1).

4. The Buntinas–Tanovic-Miller condition (see, e.g., [82]):
   Let $\{k_n\}$ be an increasing sequence and $\{m_n\}$, $1 \leq m_n \leq k_{n+1}$, a non-decreasing sequence. Then $d \in hv^p$ if

$$\sum_{n=0}^{\infty} m_n^{\frac{1}{p'}} \left\{ \sum_{k=k_n}^{k_{n+1}-1} |\Delta d_k|^p \right\}^{\frac{1}{p}} + \sum_{n=0}^{\infty} \ln \left(\frac{k_{n+1}}{m_n} \right) \sum_{k=k_n}^{k_{n+1}-1} |\Delta d_k| < \infty. \tag{3.43}$$

If $k_n = 2^n$ and $m_n = 2^{n+1}$, we get o_p. If $m_n = 1$, then

$$\sum_{n=1}^{\infty} |\Delta d_n| \ln n < \infty.$$

Buntinas and Tanovic-Miller also introduced a scale of HV^p spaces each of which is a linearization of hv^p.

5. Recently an amalgam type space (see [64, 82, 147]), in which the condition for a sequence d is

$$\sum_{n=0}^{\infty} \left\{ \sum_{m=1}^{\infty} \left[\sum_{k=m2^n}^{(m+1)2^n - 1} |\Delta d_k| \right]^2 \right\}^{\frac{1}{2}} < \infty, \tag{3.44}$$

was involved in these problems; for further analysis see [107], while for functions, analogous spaces will be discussed in the next chapter.

The traditional way to prove such conditions (or to put these sequence spaces into play) is the same application of Abel's transformation as above. This is not a surprise if we observe that all the conditions are given in terms of the differences of the coefficients (if one understands $\{d_k\}$ as one of the sequences of coefficients, either $\{a_k\}$ or $\{b_k\}$. More precisely, the absolute values of the corresponding series are integrated over $(0, \pi)$ or $(\frac{\pi}{N}, \pi)$, with appropriate bounds in terms of the coefficients. Such estimates are traditionally called *Szidon inequalities*. This name goes back to one of such inequalities obtained by Szidon [187]:

$$\frac{1}{N+1} \left\| \sum_{k=0}^{N} c_k D_k \right\|_{L^1} \le \max_{0 \le k \le N} |c_k|,$$

where c_k are arbitrary numbers. For generalizations and applications, see, e.g., a survey by Fridli [107]. Since the author of this book has never tried to prove any result for trigonometric series in this way, we shall not go into details. Instead, in Sect. 6.7 of Chap. 5 we shall return to these issues and show how the Fourier and Hilbert transforms for functions are involved in the seemingly purely sequence considerations and lead to the known and new conditions (including most of the afore-mentioned) in appropriate terms of discrete Hilbert transforms and other sequence-language notions.

Chapter 4
Fourier Transform

In this chapter, we are going to consider notions and problems similar to those in the previous chapter but in the non-compact and, correspondingly, non-periodic setting, on the real axis $\mathbb{R}$ or on its half-axis $\mathbb{R}_+ = [0, \infty)$. Here the harmonics form a continuum, and the natural function spaces decompose in a continuous way. Historically, Fourier, who, in 1811, replaced the series representation of a solution by an integral representation, and thereby initiated the study of Fourier integrals, was the pioneer of this subject. It is not by accident that the whole topic is frequently called Fourier Analysis for its discoverer.

It happens to be that I have dealt with the Fourier transform more frequently than with the Fourier series. To be precise, the Fourier transform appeared in my research from the study of the summability of Fourier series by multiplier means, where the behavior of the Fourier transform of the multiplier function plays a crucial role. Certain explanations can be found in the previous chapter. Of course, the Fourier transform lives its own life and is applicable to a diversity of problems. There is a more or less standard way to introduce the reader to the study of the Fourier transform. As usual throughout this book, we do not aim to present the basics as a complete theory (one can easily find them, in less or more detail, in any textbook); instead we emphasize certain features of the topic that first of all impressed the author earlier or later, in that or another way, and/or were systematically used (for the latter, see, first of all, Chap. 8).

4.1 Definitions and Around

We define the *Fourier transform* of g by

$$\widehat{g}(x) = \int_{-\infty}^{\infty} g(t)e^{-ixt}\,dt, \tag{4.1}$$

© The Author(s), under exclusive license to Springer Nature Switzerland AG 2021
E. Liflyand, *Harmonic Analysis on the Real Line*, Pathways in Mathematics,
https://doi.org/10.1007/978-3-030-81892-0_4

let it exist in some sense. There is no need to specify this more accurately at this point, let it be postponed unless a concrete setting is involved. We observe here that the Fourier transform is defined without any coefficient before the integral, which leads to $\frac{1}{2\pi}$ before the integral in the *inverse Fourier transform*

$$\check{h}(t) = \frac{1}{2\pi} \int_{-\infty}^{\infty} h(x)e^{ixt}\,dx. \tag{4.2}$$

One of the main questions in harmonic analysis is whether and in what sense (4.2) restores $g(t)$ if $h(x) = \widehat{g}(x)$. Or, for what functions g and at what points this is true. Frequently, the operation of taking the Fourier transform (or the Fourier coefficients in the previous chapter) is called *analysis*, as opposed to the operation of restoration of a function from its Fourier expansion, which is called *synthesis*.

The following fact was important in the study of the Fourier series and is also important for the study of the Fourier transform. It is known as the *Riemann–Lebesgue lemma* in any setting.

Lemma 4.3 *If $g \in L^1(\mathbb{R})$, then* $\lim\limits_{|x| \to \infty} \widehat{g}(x) = 0$.

Let us outline how the Riemann–Lebesgue lemma can be proved. It is immediate for the indicator function of an interval and, similarly, for the indicator function of a finite union of intervals. Then, approximating the given function f by a finite linear combination of the indicators of intervals, that is, by step functions, in the L^1 norm, we let the Fourier transform $\widehat{f}$ become small, since the error of approximation is small as well as is the Fourier transform of the approximant.

In addition, the following simple conclusion is of tremendous importance.

Lemma 4.4 *The Fourier transform of an integrable function g is uniformly continuous on $\mathbb{R}$.*

Proof Given $\varepsilon > 0$, we are going to estimate $\widehat{g}(x) - \widehat{g}(y)$. First, we find A such that

$$\int_{|t| > A} |g(t)|\,dt < \frac{\varepsilon}{4}.$$

Let now

$$|x - y| < \delta = \frac{\varepsilon}{2A\|g\|_{L^1(\mathbb{R})}}.$$

Hence,

$$|\widehat{g}(x) - \widehat{g}(y)| \leq 2\int_{|t|>A} |g(t)|\,dt + \left| \int_{|t|\leq A} g(t)[e^{-ixt} - e^{-iyt}]\,dt \right|$$

$$\leq \frac{\varepsilon}{2} + \int_{|t|\leq A} |tg(t)|\,|x - y|\,dt < \varepsilon.$$

Since for any $\varepsilon > 0$, we make $|\widehat{g}(x) - \widehat{g}(y)| < \varepsilon$ whenever $|x - y| < \delta$, with δ independent of x and y, the proof is complete. $\qquad\square$

It is reasonable to ask at this point how slow or how fast $\widehat{g}(x)$ goes to 0 as $|x| \to \infty$ provided $g \in L^1(\mathbb{R})$ (and it does go, due to the Riemann–Lebesgue lemma). The answer is as one pleases! It is one of the important features of the Fourier transform that an arbitrary (slow) rate of convergence to 0 is possible. Indeed, a classical result due to Pólya (see, e.g., [35] or [55, Chapter 6, 6.3.7]) says that each even, bounded and convex function on $[0, \infty)$ monotone decreasing to zero is the Fourier transform of an integrable function. More sophisticated examples of that kind can be found in [161]. In other words, nothing but uniform continuity and the Riemann–Lebesgue lemma can be said about the Fourier transform of an integrable function. However, we have seen that in the study of the Fourier series posing additional assumptions on the function enforce the coefficients decay faster (see, e.g., Proposition 3.14 or Sect. 3.3 in the previous chapter). Of course, the same is the case for the Fourier transform; for recent very developed results of this kind in terms of moduli of smoothness, see [115].

Very often, the cosine Fourier transform

$$\widehat{f}_c(x) = \int_0^\infty f(t) \cos xt \, dt \qquad (4.5)$$

and the sine Fourier transform

$$\widehat{f}_s(x) = \int_0^\infty f(t) \sin xt \, dt, \qquad (4.6)$$

and their integrability properties are studied rather than the general Fourier transform (4.1) and its features. In this case, the results for the general Fourier transform mostly are consequences of the obtained results for (4.5) and (4.6). Our eternal question will be whether the Fourier transform is integrable or not. In fact, a related question can be asked:

Given f uniformly continuous and vanishing at infinity, is it representable as the Fourier integral of an integrable function g, written

$$f(t) = \int_\mathbb{R} g(x)e^{itx} dx? \qquad (4.7)$$

It is said in this case that f belongs to the Wiener space (algebra) $W_0(\mathbb{R})$. Of course, in many situations g can be understood, in some sense, as the Fourier transform of f and the last formula as the Fourier inversion, maybe up to a constant multiple. A comprehensive overview of these problems is given in [161]. It is worth noting that this name and notation is frequently used for a different algebra, see, e.g., [45, 100] or [17]; on the other hand, notation A is sometimes used in place of W_0.

To get a flavor of functions with integrable Fourier transform (or similarly of those belonging to W_0), we note that such a function necessarily possesses a certain smoothness. If, say, the cosine Fourier transform $\widehat{f_c}$ is integrable on $\mathbb{R}_+$, then the integrals

$$\int_\delta^{\frac{x}{2}} \frac{f(x+t) - f(x-t)}{t} \, dt \tag{4.8}$$

are uniformly bounded (for the well-known prototype for Fourier series, see [25, Chapter II, §10], while in [55, 3.5.5] an even more subtle result of this type is given). Indeed, expressing $f(x+t)$ and $f(x-t)$ via the Fourier inversion, we obtain

$$\left| \int_\delta^{\frac{x}{2}} \frac{f(x+t) - f(x-t)}{t} \, dt \right|$$

$$= \frac{1}{\pi} \left| \int_\delta^{\frac{x}{2}} \frac{1}{t} \int_0^\infty \widehat{f_c}(u) [\cos u(x+t) - \cos u(x-t)] \, du \, dt \right|$$

$$\leq \frac{2}{\pi} \int_0^\infty |\widehat{f_c}(u)| \left| \int_\delta^{\frac{x}{2}} \frac{\sin t}{t} \, dt \right| du.$$

The last integral on the right is uniformly bounded, which is well-known. The result and proof for the sine and general Fourier transform are the same. This also emphasizes the point that not every continuous function vanishing at infinity belongs to W_0.

Of course, it is worth mentioning that the integral in the inverse formula (4.2) exists for every t if the Fourier transform is integrable, but restores the initial function almost everywhere. This is not the case for the Fourier transforms of an integrable function in general. The latter is illustrated by Kolmogorov's famous example in [133] already mentioned in the previous chapter for the Fourier series. For the Fourier transform, it works analogously and delivers an integrable function whose inverse integral diverges *everywhere*.

For functions $f \in L^p(\mathbb{R})$, with $1 < p \leq 2$, the classical Plancherel theory suggests to understand the Fourier transform as the limit in the mean (in $L^{p'}$ norm) of the truncated integrals

$$\int_{-A}^A f(t) e^{-ixt} \, dt$$

as $A \to \infty$. "To understand" means that they converge to a unique element in $L^{p'}$, which we denote by $\widehat{f}$. This is not the case for $p > 2$: the Fourier transform in this case should be understood in a different way, say distributionally. Though even in the case $1 < p \leq 2$ the Fourier transform is defined "somewhat inconcrete", there is an a priori estimate known as the *Hausdorff-Young inequality*. It reads, with $\frac{1}{p'} = 1 - \frac{1}{p}$, as

$$\|\widehat{f}\|_{L^{p'}(\mathbb{R})} \leq C_p \|f\|_{L^p(\mathbb{R})}. \tag{4.9}$$

A sharp constant C_p is due to Beckner [67] (an earlier partial result is due to K. Babenko [65]). The *Babenko–Beckner constant* is

$$C_p = (2\pi)^{\frac{1}{p'}} \frac{p^{\frac{1}{2p}}}{(p')^{\frac{1}{2p'}}}. \tag{4.10}$$

Frequently, just the constant $(2\pi)^{\frac{1}{p'}}$ is used instead, though not precise but sufficient for most applications. In many works and textbooks the Fourier transform is defined in the form

$$\widehat{g}(x) = \int_{-\infty}^{\infty} g(t)e^{-2\pi i x t} \, dt,$$

in parallel to the Fourier series for 1-periodic functions rather than 2π-periodic. In this case, that rough constant is merely 1 instead of $(2\pi)^{\frac{1}{p'}}$ (in (4.11) below as well). We note that the Hausdorff-Young inequality is one of the possible avenues where interpolation theory comes into play. A very detailed mathematical and historical overview of various variants of the Hausdorff-Young inequality is given by Butzer in [83]. He writes that "the golden thread connecting the various extensions and generalizations is the concept of logarithmic convexity". More specifically, it can be proved by interpolating between the obvious $L^1 - L^\infty$ estimate for the Fourier transform and the $L^2 - L^2$ Plancherel one (becoming the Parseval identity)

$$\|\widehat{f}\|_{L^2(\mathbb{R})} = \sqrt{2\pi} \|f\|_{L^2(\mathbb{R})}. \tag{4.11}$$

Moreover, as in Sect. 3.3 of Chap. 3, which is devoted to the absolute convergence of the Fourier series, one can be convinced that most of the sufficient conditions for the integrability of the Fourier transform are of an interpolation nature. More precisely, as mentioned there, a concept is elaborated in [160] (even in the general multivariate case) that such conditions follow from the fact that the tested function belongs to two spaces of smoothness. The situation is, roughly speaking, as follows and is similar to that mentioned in the previous chapter for Fourier series. If one parameter of smoothness is denoted by α and the other by β, then the condition is reduced to $\alpha + \beta > 1$ (greater than dimension, in general). For example, Zygmund type conditions suppose that for functions of bounded variation, any positive smoothness $\alpha > 0$ suffices. But we already know (see Sect. 2.2 in Chap. 2) that boundedness of variation means the Lip 1 smoothness ($\beta = 1$) in the L^1 metric. These amount, as above, to

$$\alpha + \beta = \alpha + 1 > 1.$$

Not reading the previous chapter or disregarding the relevant discussion in it, one may argue that in Bernstein type conditions only one smoothness is involved, roughly speaking, $\alpha > \frac{1}{2}$. Yes and no! This is merely the case where the two kinds of smoothness coincide. With $\beta = \alpha$, we still have

$$\alpha + \beta = 2\alpha > 1$$

provided that $\alpha > \frac{1}{2}$. We only mention that in [160] many more sophisticated situations are considered to confirm this heuristic principle.

In addition to the Hausdorff-Young inequality, one more inequality of the same nature must be mentioned, the well-known *Hardy–Littlewood inequality* (see, e.g., [54, Theorem 80])

$$\left(\int_{\mathbb{R}} |x|^{p-2} |\widehat{f}(x)|^p \, dx \right)^{\frac{1}{p}} \leq C_p \|f\|_{L^p(\mathbb{R})}, \qquad 1 < p \leq 2. \qquad (4.12)$$

Consideration must be given to the constraint $1 < p \leq 2$ in both inequalities, which is sharp and essential. However, in [93], taking the left-hand side of (4.12) to be finite with either

$$P_x f = \int_0^x \widehat{f}(y) \, dy$$

or

$$Q_x f = \int_x^\infty \widehat{f}(y) \, dy$$

in place of $\widehat{f}$ (cf. (7.4) in Chap. 7), the authors obtained the corresponding analogs of (4.12) for all p. Besides (4.9) and (4.12), there is a variety of inequalities for a function and its Fourier transform united under the name *Pitt's inequalities*; they will be discussed in greater detail in Sect. 8.4 of Chap. 8.

One more possible restriction to a certain effective class of functions is that to functions of bounded variation. The whole book [34] is devoted to this instance. The next assertion justifies, in a sense, the general interest in the study of the Fourier transform just for functions of bounded variation.

Proposition 4.13 *Let $g \in L^1(\mathbb{R})$ and*

$$\int_{\mathbb{R}} \frac{|\widehat{g}(x)|}{|x|} \, dx < \infty. \tag{4.14}$$

Then g is the derivative of a function of bounded variation f which is absolutely continuous, $\lim_{|t| \to \infty} f(t) = 0$, and the Fourier transform of f is integrable.

Proof Denoting

$$f(t) = \int_{-\infty}^{t} g(u) \, du,$$

we integrate by parts as follows:

$$\int_{-\infty}^{\infty} g(t)e^{-ixt} \, dt$$

$$= \left[e^{-ixt} \int_{-\infty}^{t} g(u) \, du \right]_{-\infty}^{\infty} + ix \int_{-\infty}^{\infty} \left[\int_{-\infty}^{t} g(u) \, du \right] e^{-ixt} \, dt$$

$$= ix \int_{-\infty}^{\infty} f(t)e^{-ixt} \, dt.$$

The integrated terms vanish since, by (4.14), g must have mean zero (cf. 5.16). It follows now from (4.14) that the L^1 norm of $\widehat{f}$ is finite. $\qquad\square$

This proposition is also a kind of preamble to the future study of *Hardy's inequality* (5.27).

Of course, there are many other reasons for the Fourier transform of a function of bounded variation to be studied thoroughly. For example, the Fourier multiplier is one of the central notions in analysis. It is said that a (bounded or L^∞) function m is an $X \to Y$ Fourier multiplier if the operator defined by means of the relation

$$\widehat{M_m f}(x) = m(x)\widehat{f}(x)$$

is bounded taking X to Y. One of the important facts in the theory of multipliers is that if m is a function of bounded variation on $\mathbb{R}$, then m is a Fourier multiplier on L^p (preserves L^p) for $1 < p < \infty$ (see, e.g., [12, Corollary 3.8]).

Last but not least, the Fourier transform of a function of bounded variation is well-defined for $x \neq 0$; to see this, it suffices to integrate by parts in the Stieltjes sense. Indeed, for $f \in BV_0(\mathbb{R})$, we have

$$\widehat{f}(x) = \frac{1}{ix} \int_{\mathbb{R}} e^{-ixt} \, df(t), \tag{4.15}$$

and the right-hand side is well-defined for $x \neq 0$, because of (2.23). To distinguish the integral on the right-hand side of (4.15) from the "standard" transform, (4.15) is often called the Fourier-Stieltjes transform. It is worth mentioning that rewriting the integral in (4.15) in a way of representing f as the Fourier-Stieltjes transform of the function F of bounded variation, that is,

$$f(x) = \int_{\mathbb{R}} e^{ixt} dF(t), \qquad (4.16)$$

we arrive at a different Wiener algebra $W(\mathbb{R})$. It is of the same importance as $W_0(\mathbb{R})$, the two are closely related and both are discussed in detail in [161].

4.2 From Discussion to Calculations

The time has come for some do-it-yourself work. Let us start with a few examples.

First, if $f(t)$ is $t^{-\frac{1}{2}}$ its Fourier transforms $\widehat{f}_c$ and $\widehat{f}_s$ are identical and equal to

$$\widehat{f}_c(x) = \widehat{f}_s(x) = \sqrt{\frac{\pi}{2x}}.$$

They are integrable over $(0, 1)$ but not near infinity.

Then, if $f(t)$ is e^{-t}, we have

$$\widehat{f}_c(x) = \frac{1}{1+x^2}$$

and

$$\widehat{f}_s(x) = \frac{x}{1+x^2}.$$

In this case, $\widehat{f}_c$ is integrable over $(0, +\infty)$ while $\widehat{f}_s$ is integrable only on finite intervals.

Finally, if $f(t)$ is t^{-1}, we obtain $\widehat{f}_s(x) = \frac{\pi}{2}$ for all $x > 0$ while $\widehat{f}_c(x)$ merely does not exist as an improper integral.

These examples give us a general idea of what to expect near infinity and near the origin and can be found in [5].

The results we give below were probably a sort of folklore and have never appeared till recently in an accurate form, see [163] and [34]. Clear hints are dropped by trigonometric series with monotone coefficients. For applications, we are interested in bounded functions but some of our results are valid for functions with singularities at the origin. Our functions may be not integrable on the whole half-axis $\mathbb{R}_+ = [0, \infty)$, hence the integrals are understood as improper integrals.

Theorem 4.17 *For f locally absolutely continuous on $(0, +\infty)$, vanishing at infinity: $\lim_{t \to \infty} f(t) = 0$, and monotone,*

$$\int_0^\pi |\widehat{f_c}(x)| \, dx \le \pi \int_0^1 |f(t)| \, dt + 3 \int_1^\infty \frac{|f(t)|}{t} \, dt \qquad (4.18)$$

and

$$\int_0^1 t |f(t)| \, dt + \frac{1}{12} \int_1^\infty \frac{|f(t)|}{t} \, dt$$

$$\le \int_0^\pi |\widehat{f_s}(x)| \, dx \qquad (4.19)$$

$$\le \frac{\pi^2}{2} \int_0^1 t |f(t)| \, dt + 2 \int_1^\infty \frac{|f(t)|}{t} \, dt.$$

Proof Let us begin with the cosine transform. We have

$$\int_0^\infty f(t) \cos xt \, dt = \int_0^{\frac{\pi}{x}} f(t) \cos xt \, dt - \frac{1}{x} \int_{\frac{\pi}{x}}^\infty f'(t) \sin xt \, dt. \qquad (4.20)$$

Integrating the absolute value of the first integral on the right-hand side, we obtain the estimate

$$\int_0^\pi \left| \int_0^{\frac{\pi}{x}} f(t) \cos xt \, dt \right| dx$$

$$\le \int_0^1 |f(t)| \int_0^\pi |\cos xt| \, dx \, dt + \int_1^\infty |f(t)| \int_0^{\frac{\pi}{t}} |\cos xt| \, dx \, dt$$

$$\le \pi \int_0^1 |f(t)| \, dt + 2 \int_1^\infty \frac{|f(t)|}{t} \, dt. \qquad (4.21)$$

In the second integral on the right-hand side of (4.20), we merely use the monotonicity of f and rough estimates. By this we arrive at (4.18).

It is clear that among the above examples only $f(t) = t^{-1}$ does not satisfy (4.18) in the sense that the right-hand side of (4.18) is not finite for this specific function.

For the sine transform, $\widehat{f_s}$ is of the same sign as f is. This can be found in, e.g., [54, 6.10, Theorem 123]; however, for the sake of completeness, let us prove this nice property here. Let, for simplicity, f be monotone decreasing, that is, positive. Given $x > 0$, we have

$$\int_0^\infty f(t) \sin xt \, dt = \sum_{k=0}^\infty \left[\int_{\frac{2k\pi}{x}}^{\frac{(2k+1)\pi}{x}} + \int_{\frac{(2k+1)\pi}{x}}^{\frac{(2k+2)\pi}{x}} \right] f(t) \sin xt \, dt.$$

Substituting $t \to t + \frac{\pi}{x}$ in the second integral on the right-hand side, we obtain

$$\int_0^\infty f(t) \sin xt \, dt = \sum_{k=0}^\infty \int_{\frac{2k\pi}{x}}^{\frac{(2k+1)\pi}{x}} \left[f(t) - f\left(t + \frac{\pi}{x}\right) \right] \sin xt \, dt.$$

The difference in the brackets is positive due to the monotonicity of f, also $\sin xt$ is positive on $\left(\dfrac{2k\pi}{x}, \dfrac{(2k+1)\pi}{x} \right)$, and we are done.

Again, let, for simplicity, f be monotone decreasing with, consequently, non-negative Fourier transform. We have

$$\int_0^\pi |\widehat{f_s}(x)| \, dx = \int_0^\pi \int_0^\infty f(t) \sin xt \, dt \, dx$$

$$= \int_0^2 f(t) \, dt \int_0^\pi \sin xt \, dt + \int_2^\infty f(t) \, dt \int_0^{\frac{2\pi}{t}} \sin xt \, dt$$

$$+ \int_2^\infty f(t) \, dt \int_{\frac{2\pi}{t}}^\pi \sin xt \, dt$$

$$= 2 \int_0^\infty \frac{f(t)}{t} \sin^2 \frac{\pi t}{2} \, dt. \tag{4.22}$$

The right-hand side of (4.22) is

$$2 \left(\int_0^1 + \int_1^\infty \right) \frac{f(t)}{t} \sin^2 \frac{\pi t}{2} \, dt \le \frac{\pi^2}{2} \int_0^1 t g(t) \, dt + 2 \int_1^\infty \frac{f(t)}{t} \, dt.$$

The estimate from below is derived as follows:

$$\int_0^\pi |\widehat{f_s}(x)| \, dx \ge 2 \int_0^1 \frac{f(t)}{t} \sin^2 \frac{\pi t}{2} \, dt$$

$$+ 2 \int_{\frac{5}{2}}^\infty \frac{f(t)}{t} \sin^2 \frac{\pi t}{2} \, dt$$

$$\ge 2 \int_0^1 t f(t) \, dt + 2 \sum_{k=1}^\infty \int_{2k+\frac{1}{2}}^{2k+1} \frac{f(t)}{t} \sin^2 \frac{\pi t}{2} \, dt.$$

To estimate the sum on the right, we observe that on each $(2k + \frac{1}{2}, 2k + 1)$ there holds

$$\sin^2 \frac{\pi t}{2} = \sin^2 \frac{(t - 2k)\pi}{2} \ge \frac{1}{2}$$

and, by the monotonicity of $\frac{f(t)}{t}$, we get the estimate

$$3 \int_{2k+\frac{1}{2}}^{2k+1} \frac{f(t)}{t} \, dt \geq \int_{2k+1}^{2k+2+\frac{1}{2}} \frac{f(t)}{t} \, dt. \tag{4.23}$$

These estimates yield

$$\int_0^\pi |\widehat{f_s}(x)| \, dx \geq 2 \int_0^1 t f(t) \, dt + \frac{1}{6} \int_{\frac{5}{2}}^\infty \frac{f(t)}{t} \, dt.$$

Further, using again (4.23) with $k = 0$, we obtain

$$\int_0^1 t f(t) \, dt \geq \int_{\frac{1}{2}}^1 t^2 \frac{f(t)}{t} \, dt \geq \frac{1}{12} \int_1^{\frac{5}{2}} \frac{f(t)}{t} \, dt.$$

This leads to the lower bound in (4.19), and the proof is complete.   □

We have considered integrability conditions on $[0, \pi]$. One of the reasons is that the Fourier transform of a monotone function is not necessarily integrable on the whole half-axis, see [54, 6.11, Theorem 125] or Sect. 8.5.2 in Chap. 8.

In [163], the classes of so-called general monotone functions are introduced as an analog of the same notion for sequences earlier introduced by Tikhonov in [193] (for a comprehensive survey of these classes, see [165]). The study of the L^1 integrability of the Fourier transform for functions in such classes is closely related, also by methods, to the study of the Fourier transforms for functions of bounded variation. The pattern differs greatly when studying the L^p integrability there; more about this will be presented in Sect. 8.4 of Chap. 8.

4.3   Poisson Summation Formula

One of the important tools in harmonic analysis is the Poisson summation formula. It shows how the periodic and non-periodic cases are related to one another. It was discovered by Siméon Denis Poisson and is sometimes called *Poisson resummation*.

Theorem 4.24 *If $f \in L^1(\mathbb{R})$, then the series $\sum f(x + 2k\pi)$ converges absolutely a.e. to the 2π-periodic locally integrable function whose Fourier series is of the form*

$$\sum_{k=-\infty}^\infty f(x + 2k\pi) \sim \sum_{k=-\infty}^\infty \frac{1}{2\pi} \widehat{f}(k) e^{ikx}. \tag{4.25}$$

Equality, rather than just ∼, *takes place provided both series converge everywhere and the sum of the series on the left is continuous.*

Proof By the Beppo Levi theorem, it suffices to check the convergence of the series

$$\sum_{k=-\infty}^{\infty} \int_{-\pi}^{\pi} |f(x + 2k\pi)|\, dx.$$

This is so since we have

$$\sum_{k=-\infty}^{\infty} \int_{-\pi}^{\pi} |f(x + 2k\pi)|\, dx = \sum_{k=-\infty}^{\infty} \int_{(2k-1)\pi}^{(2k+1)\pi} |f(x)|\, dx$$

$$= \int_{-\infty}^{\infty} |f(x)|\, dx = \|f\|_{L^1(\mathbb{R})} < \infty.$$

Further, the m-th Fourier coefficient, $m \in \mathbb{Z}$, equals

$$\frac{1}{2\pi} \int_{-\pi}^{\pi} \sum_{k=-\infty}^{\infty} f(x + 2k\pi) e^{-imx}\, dx$$

$$= \frac{1}{2\pi} \sum_{k=-\infty}^{\infty} \int_{-\pi}^{\pi} f(x + 2k\pi) e^{-imx}\, dx$$

$$= \frac{1}{2\pi} \sum_{k=-\infty}^{\infty} \int_{(2k-1)\pi}^{(2k+1)\pi} f(x) e^{-imx}\, dx$$

$$= \frac{1}{2\pi} \int_{-\infty}^{\infty} f(x) e^{-imx}\, dx = \frac{1}{2\pi} \widehat{f}(m).$$

It remains to take into account that the sum of the convergent Fourier series of a continuous function is just this function, which follows, for example, from the summation by the Féjer means (see Sect. 3.5 in the previous chapter). The proof is complete.   □

We note that equality holds if, for example, the function f is of bounded variation and continuous. Also, equality in (4.25) holds at every x provided f is not only integrable but also is absolutely continuous and its derivative is integrable as well (see [38, Chapter 10, §6, Exercise 2)]). One more condition for this is given in [38, Chapter 10, §6, Theorem 1]: f is continuous and both f and $\widehat{f}$ are $O(|x|^{-p})$, as $|x| \to \infty$, for some $p > 1$.

Good sources for various versions of the Poisson summation formula and their applications are [55, Chapter 3, 3.1.11] and [38, Chapter 10, §6]. Moreover, results are known which show that the Poisson summation characterizes, in some sense, the Fourier transform (see [88] and [96]).

4.4 Amalgam Type Spaces

I became interested in amalgam spaces not in a direct way. As mentioned in Sect. 3.6 of the previous chapter, an amalgam type sequence space proved to be an important subspace of bv in the study of trigonometric series. My attempts to extend this object to the function setting sparked my interest in the theory of amalgamated spaces and resulted in my introducing the notion of amalgam type function spaces in [147]. They turned out to be an interference of underlying usual amalgam spaces, and their close relation to the "genuine" amalgam spaces convinced me to give a brief presentation of the latter in this section. For the time being, let us confine ourselves to very general comments. Amalgam spaces were introduced in Wiener's work [198]. These spaces and various generalizations have become a standard notion that have a growing popularity and are being used in a variety of problems in analysis and applications. In the absence of appropriate textbooks, the reader can be referred to certain survey papers; see [101, 106, 119] to mention but a few.

However, in connection with the Fourier transform, we prefer to base ourselves on the shorter presentation in [122], where those features that seem important for a study of general harmonic analysis on $\mathbb{R}$ are highlighted. The study of what we call *amalgam type spaces* can be seen as one more way of applying standard amalgam spaces, though not by means of new spaces being amalgamated or amalgamating some others but by means of combining them in a special manner. We thus start with the definition of these classes.

Definition 4.26 We say that a locally integrable function g defined on $\mathbb{R}_+$ belongs to $A_{1,p}$ if

$$\|g\|_{A_{1,p}} = \sum_{m=-\infty}^{\infty} \left\{ \sum_{j=1}^{\infty} \left[\int_{j2^m}^{(j+1)2^m} |g(t)|\, dt \right]^p \right\}^{\frac{1}{p}} dx < \infty. \quad (4.27)$$

This space is amalgam in nature, since each of the m-th summands is the norm in the Wiener amalgam space $W(L^1, \ell^p)$ for functions

$$G_m(t) = \begin{cases} 2^m g(2^m t), & if \quad t \geq 1, \\ 0, & \text{otherwise,} \end{cases}$$

where ℓ^p, $1 \leq p < \infty$, is a space of sequences $\{d_j\}$ endowed with the norm

$$\|\{d_j\}\|_{\ell^p} = \left(\sum_{j=-\infty}^{\infty} |d_j|^p \right)^{\frac{1}{p}}. \quad (4.28)$$

The norm of a function $h : \mathbb{R} \to \mathbb{C}$ in the amalgam space $W(L^1, \ell^p)$ is taken as (see, e.g., [101, 119])

$$\left\| \left\{ \int_j^{j+1} |g(t)| \, dt \right\} \right\|_{\ell^p}.$$

In other words, we can rewrite (4.27) as

$$\|g\|_{\Lambda_{1,p}} = \sum_{m=-\infty}^{\infty} \|G_m\|_{W(L^1, \ell^p)} < \infty,$$

the sum of the norms in the Wiener type amalgam space $W(L^1, \ell^p)$ for functions $2^m g(2^m t)$ on $[2^m, \infty)$ and zero otherwise.

For the problems of integrability of the Fourier transform, $A_{1,2}$ is the most important, see [147] and [34]. It is essential that this space be a subspace of L^1. Indeed, this follows from

$$\|g\|_{A_{1,2}} \geq \sum_{m=-\infty}^{\infty} \int_{2^m}^{2^{m+1}} |g(t)| \, dt = \|g\|_{L^1(\mathbb{R}_+)}. \tag{4.29}$$

In general, amalgam spaces can be defined for a space E being a Cartesian product of a family $\{E_j : j \in \mathbb{Z}\}$ of normed spaces and for a sequence space ω_j, that is, the norm of each of the elements of E is an element of ω_j. We are not going to carry it too far, we will restrict ourselves to $E_j = L^p(j, j+1)$, $j \in \mathbb{Z}$, and $\{\omega_j\} = \ell^q$, with $1 \leq p, q \leq \infty$. By this, the amalgams are the sets (L^p, ℓ^q) of functions f (equivalence classes, in fact) on the real line $\mathbb{R}$ that are locally p-th power summable and such that

$$\|f\|_{W(L^p, \ell^q)} = \left(\sum_{j=-\infty}^{\infty} \left[\int_j^{j+1} |f(t)|^p \, dt \right]^{\frac{q}{p}} \right)^{\frac{1}{q}} \, dx < \infty, \tag{4.30}$$

with the usual conventions when p or q is infinite.

We add certain basic properties of amalgam spaces. Under this norm, the amalgam spaces $W(L^p, \ell^q)$ are Banach spaces for the indicated range of p and q. For $1 \leq p, q < \infty$, the duality relations are deduced from those for L^p and ℓ^q:

$$W(L^p, \ell^q)^* = W(L^{p'}, \ell^{q'}),$$

where p' and q' means standard conjugation. Also, the dual of $W(L^p, c_0)$, the space of functions f in $W(L^p, \ell^\infty)$ for which

$$\lim_{|j| \to \infty} \int_j^{j+1} |f(t)|^p \, dt = 0,$$

is $W(L^p, \ell^1)$. A special study of the dual of a $W(L^\infty, \ell^q)$ can be found in [122]. We only mention that $W(C_0, \ell^1)$ was introduced by N. Wiener and proved to be of great importance. The following theorem generalizes both the Plancherel theorem and the Hausdorff-Young inequality to amalgam spaces.

Theorem 4.31 *Let* $1 \le p, q \le 2$. *If* $f \in W(L^p, \ell^q)$, *then*

$$\int_{-A}^{A} f(t)e^{-ixt}\, dt$$

converges, as $A \to \infty$, *to a unique element in* $W(L^{q'}, \ell^{p'})$, *which we denote by* $\widehat{f}$. *There is a constant* $C_{p,q}$ *such that*

$$\|\widehat{f}\|_{W(L^{q'}, \ell^{p'})} \le C_{p,q} \|f\|_{W(L^p, \ell^q)}.$$

Further, if $f \in W(L^p, \ell^q)$ *and* $g \in W(L^q, \ell^p)$, *then*

$$\int_{\mathbb{R}} g(x)\widehat{f}(x)\, dx = \int_{\mathbb{R}} f(x)\widehat{g}(x)\, dx.$$

Reading up on the subject, one can see that amalgam spaces are more and becoming more involved in various studies in harmonic analysis.

4.5   Summability

Of course, the idea of summability, so natural for Fourier series, is relevant in the theory of Fourier transforms as well. The approach is very similar. Considering the Fourier transform $\widehat{f}$ of $f \in L^1(\mathbb{R})$, we cannot hope that, in general, the inverse formula

$$\frac{1}{2\pi} \int_{-\infty}^{\infty} \widehat{f}(x)e^{ixt}\, dt$$

restores $f(t)$ pointwise. Indeed, the mentioned Kolmogorov's example works here as well and delivers a function for which such an inverse formula diverges at each point. Therefore, we insert a family of multiplier functions in this integral. Mostly, such a family is generated by a single function λ in the following way:

$$\frac{1}{2\pi} \int_{-\infty}^{\infty} \lambda\left(\frac{x}{N}\right) \widehat{f}(x)e^{ixt}\, dt. \tag{4.32}$$

Now, for appropriate λ, we can hope that certain limiting processes, as $N \to \infty$, restore the initial function. As in the case of Fourier series, such a multiplier is "appropriate" if its Fourier transform behaves properly, say integrable (in symbols $\widehat{\lambda} \in L^1(\mathbb{R})$). Indeed, if one can change the order of integration in (4.32) and the Fourier transform of λ is well-defined, (4.32) can be converted into one of the following convolution type integrals:

$$\frac{N}{2\pi} \int_{-\infty}^{\infty} f(u)\widehat{\lambda}(N(u-t))\,du = \frac{1}{2\pi} \int_{-\infty}^{\infty} f\left(t + \frac{u}{N}\right)\widehat{\lambda}(u)\,du.$$

Very often, say in approximation problems, one of these forms is initially taken to approximate f.

In harmonic analysis, this process often comes into play under the name "approximation of the identity" (see, e.g., [12, Chapter 2, §1]). On the functions f from the Schwartz class $\mathcal{S} := \mathcal{S}(\mathbb{R})$, the class of infinitely differentiable functions with derivatives rapidly decreasing at infinity (faster than any polynomial), it is constructed as follows. Let φ be an integrable function on the real axis such that $\int_{\mathbb{R}} \varphi(t)\,dt = 1$. Then, denoting $\varphi_N(t) = N\varphi(Nt)$, we get that φ_N converges, as $N \to \infty$, in the distributional sense to the delta function δ. In formulas, for $f \in \mathcal{S}$, we have

$$\varphi_N(f) = \int_{\mathbb{R}} N\varphi(Nt)f(t)\,dt = \int_{\mathbb{R}} \varphi(t)f\left(\frac{t}{N}\right)dt.$$

By the Lebesgue dominated convergence theorem, we get the mentioned convergence

$$\lim_{N\to\infty} \varphi_N(f) = f(0) = \delta(f).$$

Since $\delta * f = f$, for $f \in \mathcal{S}$, we have the pointwise limit

$$\lim_{N\to\infty} \varphi_N * f(t) = f(t).$$

This argument allows one to say that the family $\{\varphi_N : N > 0\}$ is an approximation of the identity. As usual, standard machinery is then applied to extend the range of approximation of the identity to more practiced spaces, like L^p. This works more or less immediately for the convergence in norm. Most of the classical summability methods (where the multipliers were denoted by λ) can pass for approximation of the identity. For example, for Cesàro summability, both $\lambda(t)$ and $\varphi(t)$—depending on the setting—are $(1 - |t|)_+$.

Frequently, the results follow from the knowledge of behavior of Mf, the Hardy–Littlewood maximal function of a locally integrable function f. Its relation to the approximation of the identity can be seen in the following statement (see [12, Chapter 2, Proposition 2.7]).

Proposition 4.33 *Let φ be a positive even function, decreasing on $(0, \infty)$ and integrable. Then*

$$\sup_{\delta > 0} |\varphi_{\frac{1}{\delta}} * f(x)| \leq \|\varphi\|_{L^1(\mathbb{R})} Mf(x).$$

This statement is clear if φ is a simple function. But every integrable φ can be approximated by a sequence of simple functions which increase to it monotonically.

4.5.1 Summability and Poisson Summation

Back to the related problems of summability for Fourier series. They are connected to the Fourier transform via the Poisson summation formula. We return to the summability means (3.30):

$$\Lambda_N(t) = \Lambda_N(f; t) = \sum_{k=-\infty}^{\infty} \lambda\left(\frac{k}{N}\right) c_k e^{ikt}.$$

For simplicity, let us consider an integrable and continuous function λ that decays so rapidly at infinity that

$$\sum_{k=-\infty}^{\infty} |\lambda\left(\frac{k}{N}\right)| < \infty$$

for every $N > 0$. This is true for sure if λ is compactly supported. We are going to add formality to formulas like (3.34):

$$\Lambda_N = \frac{1}{2\pi} \int_{-N\pi}^{N\pi} |\widehat{\lambda}(s)| \, ds + \text{possible remainder terms.}$$

To study its behavior as $N \to \infty$, we observe that $\Lambda_N(f; \cdot) = f * K_N$, where

$$K_N(x) = \frac{1}{2\pi} \sum_{k=-\infty}^{\infty} \lambda\left(\frac{k}{N}\right) e^{ikx}.$$

The question is whether K_N is a periodic approximation of the identity. We are going to derive a simple and natural sufficient condition for this. More precisely, the integrability of the Fourier transform of the multiplier function is assumed, as mentioned in the previous chapter. First of all, it is more than natural in approximation to start with $\lambda(0) = 1$. We now have to check whether

$$\int_{-\pi}^{\pi} K_N(t)\, dt = 1.$$

Indeed, this is true since the above absolute convergence of the series allows one to integrate term-by-term:

$$\int_{-\pi}^{\pi} K_N(t)\, dt = \sum_{k=-\infty}^{\infty} \lambda\left(\frac{k}{N}\right) \int_{-\pi}^{\pi} e^{ikt}\, dt = \lambda(0) = 1.$$

The more restrictive inequality

$$\int_{-\pi}^{\pi} |K_N(t)|\, dt < \infty$$

also holds true. Setting $\mu(x) = \widehat{\lambda}(-x)$ and, like above, $\mu_N(x) = N\mu(xN)$, we inverse the Fourier transform by

$$\lambda\left(\frac{k}{N}\right) = \frac{1}{2\pi} \int_{\mathbb{R}} \widehat{\lambda}(x) e^{ix\frac{k}{N}}\, dx = N \int_{\mathbb{R}} \mu(-Nx) e^{ikx}\, dx = \widehat{\mu_N}(k).$$

By the Poisson summation formula,

$$K_N(x) = \frac{1}{2\pi} \sum_{k=-\infty}^{\infty} \widehat{\mu_N}(k) e^{ikx} = \sum_{k=-\infty}^{\infty} \mu_N(x + 2\pi k).$$

Therefore, $K_N \geq 0$ if $\mu \geq 0$ (or, in other words, if $\widehat{\lambda}(0) \geq 0$), while in the general case

$$\int_{-\pi}^{\pi} |K_N(t)|\, dt \leq \int_{-\pi}^{\pi} \sum_{k=-\infty}^{\infty} |\mu_N(x + 2\pi k)|\, dt$$

$$= \int_{\mathbb{R}} |\mu_N(x)|\, dx = \|\widehat{\lambda}\|_{L^1(\mathbb{R})}.$$

It turns out that the integrability of $\|\widehat{\lambda}\|_{L^1(\mathbb{R})}$ is also necessary for the boundedness of the L^1 norms of Λ_N, even without the assumption of continuity. Indeed, let

$$\int_{-\pi}^{\pi} \left| \sum_{k=-\infty}^{\infty} \lambda\left(\frac{k}{N}\right) e^{ikx} \right| dx \leq C$$

for some C. Substituting $y = Nx$, we have the same inequality to be of the form

$$\int_{-N\pi}^{N\pi} \left| \frac{1}{N} \sum_{k=-\infty}^{\infty} \lambda\left(\frac{k}{N}\right) e^{i\frac{k}{N}y} \right| dy \leq C.$$

Moreover, for any $R \leq N\pi$,

$$\int_{-R}^{R} \left| \frac{1}{N} \sum_{k=-\infty}^{\infty} \lambda\left(\frac{k}{N}\right) e^{i\frac{k}{N}y} \right| dy \leq C.$$

In fact, we integrate the absolute value of the Riemann integral sum for $\widehat{\lambda}$. Letting $N \to \infty$, we obtain

$$\int_{-R}^{R} |\widehat{\lambda}(y)| \, dy \leq C,$$

for arbitrary R. Therefore, we have

$$\int_{-\infty}^{\infty} |\widehat{\lambda}(y)| \, dy \leq C,$$

as claimed.

4.5.2 Wiener Algebras and Bounded Variation

We now wish to demonstrate how the summability tools help to establish relations between the Fourier transforms (or representations via Fourier integrals) and the boundedness of variation. These will be analogs of classical results for Fourier series (see, e.g., [4, Chapter I, §60] or [57, Chapter III, §4]). The main object of this topic are various spaces $X^\wedge$ of the Fourier integrals

$$f^\circ(t) \sim \int_{\mathbb{R}} g(x) e^{ixt} dx$$

such that g is the Fourier transform $\widehat{f}$ of $f \in X$

$$g(x) = \widehat{f}(x) = \frac{1}{2\pi} \int_{\mathbb{R}} f(t) e^{-ixt} dt.$$

Given X a normed space, we define $\|f^{\circ}\|_{X^{\wedge}} = \|f\|_{X}$. Recall that the notion of the Fourier-Stieltjes transform is introduced earlier. A somewhat different notation will be convenient for us in this subsection:

$$\widehat{dF}(x) = \frac{1}{2\pi} \int_{\mathbb{R}} e^{-ixt} dF(t).$$

We study conditions under which f° belongs to the space $d^{\alpha} V^{\wedge}$, $\alpha \geq 0$, that is, when

$$(ix)^{1-\alpha} g(x) = \widehat{dF}(x), \tag{4.34}$$

where similarly F is a function of bounded variation. For this, we need a specific notion of fractional derivative. The one corresponding to our scope is naturally defined via the Fourier transform: $h^{(\alpha)}$, the α-th derivative of h, is the function for which

$$\widehat{h^{(\alpha)}}(x) = (ix)^{\alpha} \widehat{h}(x).$$

In other words, belonging of f° to $d^{\alpha} V^{\wedge}$ means

$$g(x) = (ix)^{\alpha} \widehat{F}(x) = \widehat{F^{(\alpha)}}(x),$$

that is, $g(x)$ for almost all x, is the Fourier transform of the αth derivative of a function of bounded variation.

We shall study these spaces in connection with summability. Let the summability method be defined by a single function λ, a multiplier, like in (3.30), as follows:

$$\Lambda_{N}(f; x) = \int_{\mathbb{R}} \lambda\left(\frac{t}{N}\right) g(t) e^{ixt} dt.$$

It is clear that λ should be defined at each point, so let λ be continuous. The following representation is useful in many cases (when $f, \lambda \in L^{1}(\mathbb{R})$ it is merely equivalent to (3.30), see, e.g., [51, Chapter I, Theorem 1.16]; moreover, this is true under any assumptions which provide the validity of the Parseval identity)

$$\Lambda_{N}(f; x) = \int_{\mathbb{R}} N\widehat{\lambda}(N(t - x)) f(t) dt. \tag{4.35}$$

In what follows we assume that

$$\text{both } \lambda \text{ and } \widehat{\lambda} \text{ are integrable on } \mathbb{R} \tag{4.36}$$

and

$$\lambda(0) = 1. \tag{4.37}$$

We shall make use of either (3.30) or (4.35) as a definition of the linear means Λ_N just according to which of the two formulas is valid.

When a function is already represented by a Fourier integral, to take its α-th derivative means to multiply the integrand by $(it)^\alpha$. Therefore, we understand $\Lambda_N^{(1-\alpha)}$ via the representation

$$\Lambda_N^{(1-\alpha)}(f; x) = \int_{\mathbb{R}} \lambda\left(\frac{t}{N}\right) g(t) (it)^{(1-\alpha)} e^{ixt} dt.$$

The following result was proved in [66].

Theorem 4.38 Let g and λ be such that $\lambda\left(\frac{t}{N}\right) g(t) (it)^{(1-\alpha)}$ are integrable for all N. In order that f° belong to $d^\alpha V^\wedge$, it is necessary and sufficient that

$$\|\Lambda_N^{(1-\alpha)}\|_{L^1(\mathbb{R})} = O(1). \tag{4.39}$$

Remark 4.40 Clearly, $\alpha = 0$ and $\alpha = 1$ are two main cases. They are definitely related to the classes W and W_0, respectively. Slightly less general versions of these cases are known from [92] and [89]. For more information, see [10] or [26].

Proof *Necessity.*

Let $f^\circ \in d^\alpha V^\wedge$. Then

$$\Lambda_N^{(1-\alpha)}(f; x) = \int_{\mathbb{R}} \lambda\left(\frac{t}{N}\right) (it)(it)^{-\alpha} g(t) e^{ixt} dt$$

$$= \int_{\mathbb{R}} \lambda\left(\frac{t}{N}\right) \widehat{dF}(t) e^{ixt} dt.$$

By the Stieltjes version of (4.35) (see, e.g., [10, Theorem 6.5.6]), we have

$$\Lambda_N^{(1-\alpha)}(f; x) = \int_{\mathbb{R}} N\widehat{\lambda}(N(t - x)) \, dF(t).$$

Hence

$$\|\Lambda_N^{(1-\alpha)}\|_{L^1(\mathbb{R})} \leq \int_{\mathbb{R}} \int_{\mathbb{R}} N\left|\widehat{\lambda}(N(t - x))\right| |dF(t)| \, dx.$$

Since $\widehat{\lambda}$ is integrable on $\mathbb{R}$, and F is of bounded variation we may use here and in what follows the Fubini theorem freely. This yields

$$\|\Lambda_N^{(1-\alpha)}\|_{L^1(\mathbb{R})} \le \int_{\mathbb{R}} |dF(t)| \int_{\mathbb{R}} N|\widehat{\lambda}(N(t-x))|\,dx$$

$$= \|F\|_{BV}\|\widehat{\lambda}\|_{L^1(\mathbb{R})},$$

and we are done.

Sufficiency.

Denote

$$\Phi_N(x) = \int_{-\infty}^{x} \Lambda_N^{(1-\alpha)}(f;t)\,dt.$$

This sequence possesses the following properties. First, (4.39) provides that both the family $\{\Phi_N\}$ and the family of variations of these functions are uniformly bounded (by the same constant $\|\Lambda_N^{(1-\alpha)}\|_{L^1(\mathbb{R})}$). In virtue of the boundedness of $\{\Phi_N\}$ along with their variations, Helly's first theorem (see, e.g., [26, Theorem 9.1.1] or Sect. 2.2) ensures the existence of a subsequence $\{N_k\}$, $N_k \to \infty$ as $k \to \infty$, such that

$$\lim_{k\to\infty} \Phi_{N_k} = \Phi(x)$$

at any point x on the whole $\mathbb{R}$, where Φ is a function of bounded variation (bounded by $\|\Lambda_N^{(1-\alpha)}\|_{L^1(\mathbb{R})}$ along with its total variation). Our next (and final) step is to show that

$$\widehat{d\Phi}(x) = \frac{1}{2\pi}\int_{\mathbb{R}} e^{-ixt}d\Phi(t) = \lim_{k\to\infty}\frac{1}{2\pi}\int_{\mathbb{R}} e^{-ixt}d\Phi_{N_k}(t). \qquad (4.41)$$

Generally speaking, Helly's second theorem (see Theorem 2.25 in Sect. 2.2) is true only under additional assumptions (see, e.g., [26, Theorem 9.1.3]). For example, it holds true if the last limit is uniform on every finite interval (see [77, Lemma 1]). Similarly to [89], we have

$$\lambda\left(\frac{t}{N_k}\right)(it)^{1-\alpha}g(t) = \frac{1}{2\pi}\int_{\mathbb{R}} e^{-ixt}\,d\Phi_{N_k}(x).$$

The right-hand side is continuous as well as $\lambda\left(\frac{t}{N_k}\right)$, hence $(it)^{1-\alpha}g(t)$ almost everywhere coincides with a continuous function $\psi(t)$. Now, by (4.36) and (4.37), we get that $\lambda\left(\frac{t}{N_k}\right)\psi(t)$ converges, as $N \to \infty$, to $\psi(t)$ uniformly on every finite interval. Thus, (4.41) is true, which completes the proof. □

Taking the Cesàro means as summability method, we obtain a transparent condition for a function in W_0 to be of bounded variation. It is a direct corollary of the above theorem. We will present it in terms of (4.7).

Corollary 4.42 *For $f \in W_0(\mathbb{R})$ to be of bounded variation, it is necessary and sufficient that*

$$\sup_{N>0} \int_{\mathbb{R}} \left| \int_{-N}^{N} t\left(1 - \frac{|t|}{N}\right) g(t) e^{ixt} \, dt \right| \, dx < \infty. \tag{4.43}$$

In fact, Cramér in [89] attributes this result to Hausdorff's work [118].

Chapter 5
Hilbert Transform

There are two main operators in harmonic analysis on the real line. One of them, the Fourier transform in various forms, has been discussed in the previous chapters. We now proceed to the second one, the Hilbert transform. We precede the basic presentation in a special manner. Wikipedia says that the Hilbert transform is a technique used to obtain the minimum-phase response from a spectral analysis. Pandey begins his book [43] by listing the fields where the Hilbert transform arises:

- (i) Signal processing (the Hilbert transform of periodic functions)
- (ii) Metallurgy (Griffith crack problem and the theory of elasticity)
- (iii) Dirichlet boundary value problems (potential theory)
- (iv) Dispersion relation in high energy physics, spectroscopy, and wave equations
- (v) Wing theory
- (vi) The Hilbert problem
- (vii) Harmonic analysis
    Liu in his survey [169] adds more applications:
- (viii) Sampling of bandpass signals for communication
- (ix) AM-FM decomposition for auditory prostheses
- (x) System identification

In the sequel, both above mentioned authors explain some of these applications in more detail. However, we shall deal with the Hilbert transform within the framework of harmonic analysis only.

© The Author(s), under exclusive license to Springer Nature Switzerland AG 2021   71
E. Liflyand, *Harmonic Analysis on the Real Line*, Pathways in Mathematics,
https://doi.org/10.1007/978-3-030-81892-0_5

5.1 Definitions and Calculations

The Hilbert transform of a function g is

$$\mathcal{H}g(x) = \frac{1}{\pi} \int_{\mathbb{R}} \frac{g(t)}{x - t} \, dt, \tag{5.1}$$

where the integral is understood in the improper (Principal Value) sense, as

$$\mathcal{H}g(x) = \lim_{\delta \to 0+} \mathcal{H}_\delta g(x),$$

with

$$\mathcal{H}_\delta g(x) = \frac{1}{\pi} \int_{|t-x|>\delta} \frac{g(t)}{x - t} \, dt$$

being the *truncated Hilbert transform*. We mention that the study of the maximal Hilbert transform

$$\mathcal{H}^* g(x) = \sup_{\delta > 0} |\mathcal{H}_\delta g(x)|$$

is certainly no less important. However, we shall not pay much attention to it, the necessary matter can be found in many of the referred to sources.

There is an immense amount of motivations, or, in other words, problems that motivate the use of the Hilbert transform, in relation to the Fourier transform or without explicit relation. One of them, of course, comes from complex analysis, where the Hilbert transform characterizes analytic functions in the upper half-plane. In addition to the above, recall also that it appears in the inversion formula of the Radon transform, see, e.g., [41, Chapter II]. We need not even mention the role of computer tomography in our life today. However, our main interest in the Hilbert transform and related tools is of a purely analytic nature and lies within the scope of harmonic analysis. In the next chapter, we will consider the real Hardy space $H^1(\mathbb{R})$, the subspace of $L^1(\mathbb{R})$ functions with integrable Hilbert transform. That space is frequently considered as a setting for improvement of various properties and inequalities that fail to hold in $L^1(\mathbb{R})$. For example, studying integrability of the Fourier transform of a function with bounded variation, we learn that for such functions, for which we mainly know that the derivative is integrable, the integrability of the Fourier transform may fail. However, if we know that the derivative lies in $H^1(\mathbb{R})$, the Fourier transform is integrable. For a detailed study of such problems, see [34, Part I].

We shall denote the way of integration by means of which the Hilbert transform is defined by P.V. before the integral sign, if needed. For example, (5.1) without further explanations, can be briefly written as

$$\mathcal{H}g(x) = \frac{1}{\pi} \text{P.V.} \int_{\mathbb{R}} \frac{g(t)}{x - t} \, dt.$$

Let us consider a few simple examples. Each of them is widely used in applications and textbooks.

Example 5.2 The Hilbert transform of a constant function is identically zero. This is because of the symmetry in the principal value approach.

Example 5.3 The Hilbert transform of the indicator function $\chi_{[a,b]}$ exists except at the two points a and b and is given by

$$\mathcal{H}\chi_{[a,b]}(x) = \frac{1}{\pi} \ln \frac{|x - a|}{|x - b|}. \tag{5.4}$$

Let $x \neq a, b$ and $\delta < \min\{|x - a|, |x - b|\}$. If $x - b > 0$ or $x - a < 0$, then (5.4) follows immediately by simple integration. If $a < x < b$, then

$$\mathcal{H}\chi_{[a,b]}(x) = \frac{1}{\pi} \lim_{\delta \to 0+} \left(\ln \frac{|x - a|}{\delta} + \ln \frac{\delta}{|x - b|} \right),$$

which implies (5.4). In this last case, we essentially use that the transform is understood in the principal value sense.

Example 5.5 The Hilbert transform of $g(t) = \sin at, a > 0$, on the real line is given by $\mathcal{H}g(x) = -\cos ax$.

We have

$$\lim_{\delta \to 0+} \frac{1}{\pi} \int_{|x-t| \geq \delta} \frac{\sin at}{x - t} \, dt = \frac{1}{\pi} \lim_{\delta \to 0+} \int_{|t| \geq \delta} \frac{\sin(ax - at)}{t} \, dt$$

$$= \frac{\sin ax}{\pi} \lim_{\delta \to 0+} \int_{|t| \geq \delta} \frac{\cos at}{t} \, dt - \frac{\cos ax}{\pi} \lim_{\delta \to 0+} \int_{|t| \geq \delta} \frac{\sin at}{t} \, dt.$$

The first integral on the right-hand side vanishes because of the oddness of the integrand. The second one is well known and equals π. Along with the previous factor, it gives the required result.

Example 5.6 In a similar way, the Hilbert transform of $g(t) = \cos at, a > 0$, on the real line is given by $\mathcal{H}g(x) = \sin ax$.

Because of the last two relations, engineers sometimes call the Hilbert transform (see [28, Ch. VIII, §7]) a "90° phase shift". The notion is of such importance that there is even a piece of hardware called a "Hilbert transformer" that takes an input signal and produces some kind of approximation to the Hilbert transform of the signal.

In these examples, we were able to use the definition of the Hilbert transform. However, we must have a deeper knowledge on the existence of the Hilbert transform for functions in various classes. Prior to this, we shall try to realize how the Hilbert transform may appear in general. What we mean is that there are various ways in which the Hilbert transform comes into play. Not that one of them is definitely better than the others—it is much a question of the topic and taste.

Let us mention a more general notion, the so-called *Hilbert-Stieltjes transform*:

$$\mathcal{H}df(x) = \frac{1}{\pi} \int_{\mathbb{R}} \frac{df(t)}{x - t}, \tag{5.7}$$

the Hilbert transform of the measure df generated by a function of bounded variation f subject to the condition $\lim_{t \to -\infty} f(t) = 0$. Being always understood in the principal value sense, it reduces to the usual Hilbert transform for absolutely continuous f. Many details can be found, e.g., in [10, 8.1.1]; however, let us mention, for example, that (5.7) was widely used as early as 1922 by R. Nevanlinna in [173].

5.2   The Hilbert Transform Comes into Play

In [34], we decided to reduce the "early days" of the Hilbert transform to one of the oldest and somewhat "naive", heuristic approaches in regard to rigorousness. Here we prefer to switch to a modern approach. We shall not touch in detail the aspects of complex analysis or the theory of distributions. They will be used (or, more precisely, mentioned as in Chap. 4) only, with corresponding references if needed. Among numerous sources, we again prefer the lines of [12]. However, there are many sources where the Hilbert transform is introduced more or less similarly and everyone can find one that is to one's taste (a rich bibliography can be found in [27]). In any case, we shall only give an outline of the setting.

First, we start with functions f from the Schwartz class $\mathcal{S}(\mathbb{R})$. The algorithm is as follows. Such a function can be extended harmonically to the upper half-plane by convolving with the *Poisson kernel*

$$P_t(x) = \frac{1}{\pi} \frac{t}{t^2 + x^2}.$$

This harmonic function $u(x, t) = P_t * f(x)$ is the real part of the corresponding analytic function, whose imaginary part (harmonic conjugate of u) can be written as

$$v(z) = \int_{\mathbb{R}} -i \operatorname{sign} s e^{-2\pi t |s|} \widehat{f}(s) e^{2\pi i x s} \, ds,$$

which is equivalent to

$$v(x, t) = Q_t * f(x),$$

where

$$\widehat{Q_t}(s) = -i \, \text{sign} \, s e^{-2\pi t |s|}. \tag{5.8}$$

By the Fourier inversion, we have

$$Q_t(x) = \frac{1}{\pi} \frac{x}{t^2 + x^2},$$

the *conjugate Poisson kernel*, that is, conjugate of P_t. The difference between the two kernels is that P_t is an approximation of the identity (see Sect. 4.5 in the previous chapter), while Q_t is not. Since

$$\lim_{t \to 0} Q_t(x) = \frac{1}{\pi x}$$

is not locally integrable, we cannot convolve it with smooth functions. A solution of this difficulty is as follows. Since the regular way for working with $\frac{1}{x}$ is closed, a tempered distribution is defined called the principal value of $\frac{1}{x}$:

$$\text{P.V.} \frac{1}{x} (\psi) = \lim_{\delta \to 0+} \int_{|x| > \delta} \frac{\psi(x)}{x} \, dx,$$

with $\psi \in \mathcal{S}$. It can be proved that in the topology of the space of tempered distributions $\mathcal{S}'$, there holds

$$\lim_{t \to 0} Q_t = \frac{1}{\pi} \text{P.V.} \frac{1}{x}.$$

It follows from this that $\lim_{t \to 0} Q_t * f(x)$ is exactly (5.1). By the continuity of the Fourier transform in $\mathcal{S}'$ and by (5.8), the Fourier transform of $\frac{1}{\pi} \text{P.V.} \frac{1}{x}$ at the point s is $-i \, \text{sign} \, s$. This leads to the possibility of additionally defining the Hilbert transform of $f \in \mathcal{S}$ by

$$\widehat{\mathcal{H}f}(x) = -i \, \text{sign} \, x \, \widehat{f}(x). \tag{5.9}$$

The latter allows one to define the Hilbert transform also for functions in $L^2(\mathbb{R})$, with

$$\|\mathcal{H}f\|_{L^2(\mathbb{R})} = \|f\|_{L^2(\mathbb{R})} \tag{5.10}$$

and the property $\mathcal{H}(\mathcal{H}f) = -f$.

Back to the variety of the ways to define (or, say, to introduce) the Hilbert transform, let us mention the one that heuristically comes from the Fourier integral formula, as in Chapter V of Titchmarsh's celebrated book [54]. It naturally leads to

$$\mathcal{H}g(t) = -\int_0^\infty [b(x)\cos tx - a(x)\sin tx]\,dx$$

$$= \frac{1}{\pi}\int_0^\infty \int_{-\infty}^\infty \sin(t - v)x\, g(v)\,dv\,dx. \tag{5.11}$$

The integral on the right-hand side of (5.11) is called the *conjugate integral of the Fourier integral*. In what follows, we shall demonstrate how the summability reduces this integral exactly to the Hilbert transform.

Relation (5.9) can be involved in defining the Hilbert transform in a somewhat different way (see, e.g., [20, Chapter 9, 9.3]). More precisely, via the Fourier transforms of one-sided functions, which are nonzero only on the right half-axis. Such functions represent several real phenomena. A signal that turns on at the origin or the impulse response of a causal linear system (one which does not produce an output before the input is applied) are one-sided. Causality is an important constraint on the design of systems that operate in real time on streams of data. In fact, the real and imaginary parts of the Fourier transforms of one-sided functions are not independent, but can be calculated one from the other. To see how this works, we break the one-sided function f into its even and odd parts, as in (3.4):

$$f(t) = \frac{f(t) + f(-t)}{2} + \frac{f(t) - f(-t)}{2} := f_e(t) + f_o(t).$$

Thus, the even and odd parts are connected, by

$$f_o(t) = f_e(t)\operatorname{sign} t.$$

The Fourier transform of both sides of this expression should now be taken. The Fourier transform of a real and even function is real, and the transform of a real and odd function is imaginary. We then have

$$\widehat{f} = \widehat{f_e} + i\widehat{f_o},$$

and $\widehat{f_0}$ is related to $\widehat{f_e}$ by

$$i\widehat{f_o}(x) = \widehat{f_e} * \widehat{\operatorname{sign}\cdot}(x).$$

From here we arrive at familiar formulas.

5.3 Existence Almost Everywhere

One of the most important properties of the Hilbert transform of an integrable function is its *weak estimate*. In this sense, we can compare the behavior of the Hilbert transform with that of the Hardy–Littlewood maximal function (see Sect. 2.1.3 in Chap. 2). However, there is a serious difference. The latter can never be integrable while there is a rich family of integrable functions with integrable Hilbert transform; we are going to return to this problem in detail in the following chapter. It will give us the possibility to extend the Hilbert transform, previous defined in $\mathcal{S}$ or $L^2(\mathbb{R})$ to functions in $L^p(\mathbb{R})$, $1 \leq p < \infty$. This is also one of the most natural ways to prove the existence almost everywhere of the Hilbert transform.

5.3.1 Weak Estimate

In the sequel, we shall discuss the problems of integrability of the Hilbert transform. Though the Hilbert transform of an integrable function need not be integrable, it is subject to the so-called weak estimate due to Kolmogorov. This will also lead to the fact that the Hilbert transform of an integrable function exists almost everywhere. Recall that by $|E|$ we denote the Lebesgue measure of the set E; we again hope that there will be no confusion with the absolute value notation.

Theorem 5.12 *For $f \in \mathcal{S}(\mathbb{R})$ and $\lambda > 0$, there holds*

$$|\{x \in \mathbb{R} : |\mathcal{H}f(x)| > \lambda\}| \leq \frac{C}{\lambda} \|f\|_{L^1(\mathbb{R})}. \tag{5.13}$$

Proof Since a complex-valued function has the real and imaginary parts, and a real-valued function can be decomposed into its positive and negative parts f_1 and f_2 by

$$f(t) = \frac{f(t) + |f(t)|}{2} + \frac{f(t) - |f(t)|}{2} := f_1(t) + f_2(t),$$

it suffices to assume f be non-negative. Given $f \geq 0$ and $\lambda > 0$, we form the Calderón-Zygmund decomposition as in Sect. 2.1.4 of Chap. 2. This decomposition yields a sequence of disjoint intervals $\{I_j\}$ for which

1. $f(x) \leq \lambda$ for almost every $x \notin \Omega = \bigcup_j I_j$;

2. $|\Omega| \leq \frac{1}{\lambda} \|f\|_{L^1(\mathbb{R})}$;
3. $\lambda < \frac{1}{|I_j|} \int_{I_j} f(t)\, dt \leq 2\lambda$.

We now decompose $f(x)$ as the sum of two functions

$$g(x) = \begin{cases} f(x), & x \notin \Omega, \\ \frac{1}{|I_j|} \int_{I_j} f(t)\, dt, & x \in I_j, \end{cases}$$

and

$$b(x) = \sum_j b_j(x),$$

with

$$b_j(x) = \left(f(x) - \frac{1}{|I_j|} \int_{I_j} f(t)\, dt \right) \chi_{I_j}(x).$$

These two summands are traditionally called the *good* and *bad* parts of f, respectively. We have $g(x) \leq 2\lambda$ almost everywhere. As for the components of b, we get that b_j is supported on I_j and has mean zero. Further,

$$|\{x \in \mathbb{R} : |\mathcal{H}f(x)| > \lambda\}| \leq |\{x \in \mathbb{R} : |\mathcal{H}g(x)| \geq \frac{\lambda}{2}\}|$$

$$+ |\{x \in \mathbb{R} : |\mathcal{H}b(x)| \geq \frac{\lambda}{2}\}|.$$

For the first term on the right-hand side, Chebyshev's inequality and the above L^2 norm argument (5.10) yield

$$|\{x \in \mathbb{R} : |\mathcal{H}g(x)| \geq \frac{\lambda}{2}\}| \leq \left(\frac{2}{\lambda} \right)^2 \int_{\mathbb{R}} |\mathcal{H}g(x)|^2\, dx$$

$$= \frac{4}{\lambda^2} \int_{\mathbb{R}} g(x)^2\, dx$$

$$\leq \frac{8}{\lambda} \int_{\mathbb{R}} g(x)\, dx = \frac{8}{\lambda} \int_{\mathbb{R}} f(x)\, dx.$$

We now proceed to the second summand. Let $2I_j$ be the interval with the same center c_j as I_j but twice longer, that is, the 2-homothety of I_j. Let

$$\Omega^* = 2\Omega = \bigcup_j 2I_j.$$

We then have $|\Omega^*| \leq 2|\Omega|$ and

$$|\{x \in \mathbb{R} : |\mathcal{H}b(x)| \geq \frac{\lambda}{2}\}| \leq |\Omega^*| + |\{x \notin \Omega^* : |\mathcal{H}b(x)| \geq \frac{\lambda}{2}\}|$$

$$\leq \frac{2}{\lambda}\|f\|_{L^1(\mathbb{R})} + \frac{2}{\lambda}\int_{\mathbb{R}\setminus\Omega^*} |\mathcal{H}b(x)|\,dx.$$

Further, we have

$$|\mathcal{H}b(x)| \leq \sum_j |\mathcal{H}b_j(x)|$$

almost everywhere. This is obvious if the sum is finite; if not, by (5.10), this follows from the convergence of $\sum b_j$ and $\sum \mathcal{H}b_j$ to b and $\mathcal{H}b$, respectively, in L^2. To complete the proof of the theorem, it suffices to show that

$$\sum_j \int_{\mathbb{R}\setminus 2I_j} |\mathcal{H}b_j(x)|\,dx \leq C\|f\|_{L^1(\mathbb{R})}.$$

Despite $b_j \notin S$ for $x \notin 2I_j$, the formula

$$\mathcal{H}b_j(x) = \int_{I_j} \frac{b_j(t)}{x - t}\,dt$$

is still valid. Using the mean zero property of b_j, we obtain

$$\int_{\mathbb{R}\setminus 2I_j} |\mathcal{H}b_j(x)|\,dx = \int_{\mathbb{R}\setminus 2I_j} \left| \int_{I_j} b_j(t) \left[\frac{1}{x-t} - \frac{1}{x-c_j} \right] dt \right| dx$$

$$\leq \int_{I_j} |b_j(t)| \left(\int_{\mathbb{R}\setminus 2I_j} \frac{|t - c_j|}{|x - t|\,|x - c_j|}\,dx \right) dt.$$

Since $|t - c_j| \leq \frac{|I_j|}{2}$ and $|x - t| \geq \frac{|x - c_j|}{2}$, the last integral does not exceed

$$\int_{I_j} |b_j(t)| \left(\int_{\mathbb{R}\setminus 2I_j} \frac{|I_j|}{|x - c_j|^2}\,dx \right) dt.$$

The inner integral is exactly 2; therefore

$$\sum_j \int_{\mathbb{R}\setminus 2I_j} |\mathcal{H}b_j(x)|\,dx \leq 2\sum_j \int_{I_j} |b_j(t)|\,dt \leq 4\|f\|_{L^1(\mathbb{R})},$$

which completes the proof. $\square$

A similar weak estimate is true for the Hilbert-Stieltjes transform (5.7) as well. In [34], we preferred to derive the needed properties of the Hilbert transform from those for the Hilbert-Stieltjes transform.

5.3.2 Extension to L^1

The obtained weak estimate allows us to extend the Hilbert transform to functions in L^1. Indeed, for any $f \in L^1$, there is a sequence of functions f_n in S which converges to f in L^1 norm. By the weak inequality, the sequence $\{\mathcal{H} f_n\}$ is a Cauchy sequence in measure< that is, for any $\varepsilon > 0$,

$$\lim_{m,n \to \infty} \left| \left\{ x \in \mathbb{R} : |\mathcal{H} f_n(x) - \mathcal{H} f_m(x)| > \varepsilon \right\} \right| = 0.$$

Therefore, it converges in measure to a measurable function which is defined to be the Hilbert transform $\mathcal{H} f$ of f. There is a subsequence of $\{\mathcal{H} f_n\}$, depending on f, which converges pointwise almost everywhere to the above defined $\mathcal{H} f$.

We also need to know that $\lim_{\delta \to 0+} \mathcal{H}_\delta f(x)$ exists almost everywhere. One of the standard ways to prove this is to consider

$$\left| \limsup_{\delta \to 0+} \mathcal{H}_\delta f(x) - \liminf_{\delta \to 0+} \mathcal{H}_\delta f(x) \right|.$$

It would follow from the weak estimate for the maximal Hilbert transform $\mathcal{H}^*$ that this function vanishes almost everywhere; therefore, the needed limit exists almost everywhere. However, so far, we have the weak estimate for the usual Hilbert transform $\mathcal{H}$. In fact, the weak estimate for the maximal Hilbert transform $\mathcal{H}^*$ can be proved along the same lines as that above (we thus omit this proof, see [12, Chapter 3, Theorem 3.4]) with the following Cotlar's inequality (5.15) in hand. Recall that we denote by Mf the Hardy–Littlewood maximal function of a locally integrable function f, see Sect. 2.1.3 of Chap. 2.

Lemma 5.14 *If $f \in S$, then we have*

$$\mathcal{H}^* f(x) \leq M(\mathcal{H} f)(x) + C M f(x). \tag{5.15}$$

In turn, the proof of this inequality strongly rests on Proposition 4.33 in the previous chapter.

One more way to establish the almost everywhere existence is by means of complex analysis. We have already seen the relation between the two topics.

In connection with the subject of this section, it is worth mentioning a recent result by Karagulyan in [129]: for any set of measure zero on $\mathbb{R}$, there is an integrable function whose Hilbert transform does not exist at this set. This means that we cannot hope for the existence of the Hilbert transform of *all* integrable functions at all the Lebesgue points. Nor can we hope for this on any fixed set. However, there is an interesting open problem of establishing, for a given integrable function, a set of full measure where the Hilbert transform does exist. Of course, such a set may vary from function to function. Compounding such a question and discussing it with colleagues, the author learned from Javad Mashreghi that earlier V.P. Havin suggested this problem to him.

Back to increasing rearrangements (2.4), their relations to the Hilbert transform is given by

$$(\mathcal{H}f)^*(t) \leq C \left(\frac{1}{t} \int_0^t f^*(t)\, dt + \int_t^\infty f^*(t)\, \frac{dt}{t} \right)$$

for all $t > 0$ (clearly, notation in this relation differs from that for the maximal Hilbert transform $\mathcal{H}^*$; $(\mathcal{H}f)^*$ on the left-hand side means the increasing rearrangement of the function $\mathcal{H}f$). The proof can be found in, e.g., [6]; this inequality proved to be useful for many applications.

5.4   Integrability of the Hilbert Transform

In most situations, we shall consider the Hilbert transform $\mathcal{H}g$ of an integrable function g. This choice is stipulated by the problems we are going to discuss and deal with. We have already proved that such an $\mathcal{H}g$ exists almost everywhere. However, in a similar way as above we can extend the problems of existence of the Hilbert transform almost everywhere to L^p functions, with $1 < p < \infty$. We do not concentrate on this, since by the M. Riesz (projection) theorem, the class of functions with L^p integrable Hilbert transform coincides with L^p. In other words, the Hilbert transform of an L^p function is always in L^p. We shall return to this issue in the following chapter. On the other hand, if g is integrable, its Hilbert transform $\mathcal{H}g$ is not necessarily integrable. However, when it is, g has mean zero (or, in other words, possesses the *cancelation property*)

$$\int_{\mathbb{R}} g(t)\, dt = 0. \tag{5.16}$$

This was apparently first mentioned in [130].

Let us consider a few examples, with an eye on whether the Hilbert transform is integrable or not. The first one can be found in almost every source on the subject (see, e.g., [27, p.9]), while the rest are less known.

Example 5.17 The Hilbert transform of

$$g(t) = \frac{1}{1 + t^2}$$

on the real line is given by

$$\mathcal{H}g(x) = \frac{x}{1+x^2}.$$

We have

$$\mathcal{H}g(x) = \lim_{\delta \to 0+} \frac{1}{\pi} \int_{|x-t| \geq \delta} \frac{1}{(x-t)(1+t^2)} \, dt.$$

It is easy to check that

$$(1+x^2)\frac{1}{(x-t)(1+t^2)} = \frac{1}{x-t} + \frac{t}{1+t^2} + \frac{x}{1+t^2}.$$

Integration of the first summand on the right-hand side means the Hilbert transform of a constant and thus vanishes because of Example 5.2. Integration of the second one in the principal value sense gives zero because of the oddness of the integrand. Integrating the third one leads to

$$\mathcal{H}g(x) = \frac{x}{\pi(1+x^2)} \lim_{\delta \to 0+} \int_{|x-t| \geq \delta} (\arctan t \, |_{-\infty}^{x-\delta} + \arctan t \, |_{x+\delta}^{\infty}) \, dt$$

$$= \frac{x}{1+x^2}. \tag{5.18}$$

Seeing that the Hilbert transform of such a nice function is not integrable, one may guess in a substantiated way that the only reason for this is the lack of the cancelation property (5.16).

The next example of the Hilbert transform, also non-integrable, can be found in [54, 5.14].

Example 5.19 Let

$$g(t) = \begin{cases} \frac{1}{t \ln^2 t}, & t > 0, \\ \\ 0, & t \leq 0. \end{cases}$$

Then for $x > 0$,

$$|\mathcal{H}g(-x)| = \frac{1}{\pi} \int_{\mathbb{R}} \frac{g(t)}{t+x} \, dt > \frac{1}{\pi} \int_0^x \frac{dt}{2xt \ln^2 t} = \frac{1}{2\pi x \ln x}.$$

Obviously, $\mathcal{H}g \notin L^1(\mathbb{R})$. Unlike the previous example, this function satisfies (5.16).

Below we will give a similar example with a function also satisfying (5.16) but bounded.

Example 5.20 There exists an odd function with non-integrable Hilbert transform: take

$$g(t) = \frac{1}{(t-1)|\ln^2(t-1)|}$$

on $(1, \frac{3}{2})$, $g(t) = -g(-t)$ on $(-\frac{3}{2}, -1)$, and 0 otherwise. Then, for $x \in (\frac{1}{2}, 1)$, we have

$$|\mathcal{H}g(x)| \geq \left| \int_1^{1+(1-x)} \frac{1}{(t-1)\ln^2(t-1)} \frac{dt}{t-x} \right| - \frac{2}{3\ln 2}$$

$$\geq \frac{1}{2(1-x)|\ln(1-x)|} - \frac{2}{3\ln 2},$$

which is obviously non-integrable.

This example destroys the idea that an odd integrable function, always satisfying (5.16), makes the Hilbert transform automatically integrable. Surprisingly, this example, as an example of an odd function with non-integrable Hilbert transform, explicitly appeared in the literature as recently as in 2010, see [164].

Similarly, an example in the even case is a modification of Pitt's example given in [130, Theorem 1 (b)].

Example 5.21 Take

$$g_1(t) = \frac{1}{t \ln^2 t}$$

on $[2, \infty)$ and zero otherwise, and $g_2(t) = \frac{2}{\ln 2}$ in $(0, \frac{1}{2})$ and zero otherwise. Let $g(t) = g_1(t) - g_2(t)$.

This function satisfies (5.16), is integrable on $\mathbb{R}$ and, by routine calculations as above, its Hilbert transform does not belong to $L^1(-\frac{1}{2}, 0)$. It remains to extend this function evenly and take into account that the even extension possesses the same integrability properties (see, e.g., [13, Chapter III, Lemma 7.40, p.354]).

In fact, an example can be given of an integrable function whose Hilbert transform fails to be, nay, locally integrable.

5.5   Special Cases of the Hilbert Transform

When in the definition of the Hilbert transform (5.1) the function g is odd, we will denote this transform by $\mathcal{H}_o$, and by simple substitution it is equal to

$$\mathcal{H}_o g(x) = \frac{2}{\pi} \int_0^\infty \frac{t g(t)}{x^2 - t^2}\, dt. \qquad (5.22)$$

Correspondingly, when in the definition of the Hilbert transform (5.1) the function g is even, we will denote this transform by $\mathcal{H}_e$, and in a similar way it is equal to

$$\mathcal{H}_e g(x) = \frac{2}{\pi} \int_0^\infty \frac{x g(t)}{x^2 - t^2}\, dt. \qquad (5.23)$$

For a function g defined on the half-axis, these two specific transforms are connected in the following way.

Proposition 5.24 *Let $g \in L^1(\mathbb{R}_+)$. Then*

$$\mathcal{H}_e g(x) = \mathcal{H}_o g(x) + \frac{2}{\pi x} \int_0^x g(t)\, dt + G(x), \qquad (5.25)$$

where

$$\int_0^\infty |G(x)|\, dx \lesssim \int_0^\infty |g(t)|\, dt.$$

Proof The two transforms differ from one another by the factor t or x in the nominator. They can be interchanged:

$$\mathcal{H}_e g(x) = \mathcal{H}_o g(x) + \frac{2}{\pi} \int_0^\infty \frac{g(t)}{x + t}\, dt. \qquad (5.26)$$

Rewriting the last integral as

$$\int_0^\infty \frac{g(t)}{x + t}\, dt = \frac{1}{x} \int_0^x g(t)\, dt$$

$$- \int_0^x g(t) \frac{t}{x(x + t)}\, dt + \int_x^\infty \frac{g(t)}{x + t}\, dt,$$

we can check that the last two summands on the right are integrable, as required. Both are treated by means of Fubini's theorem. For the first of these two summands, we have

$$\int_0^\infty \left| \int_0^x g(t) \frac{t}{x(x+t)} \, dt \right| dx$$

$$\leq \int_0^\infty |g(t)| \int_t^\infty \left(\frac{1}{x} - \frac{1}{x+t} \right) dx \, dt = \ln 2 \int_0^\infty |g(t)| \, dt.$$

The estimate for the second one goes along the same lines:

$$\int_0^\infty \left| \int_x^\infty g(t) \frac{1}{x+t} \, dt \right| dx \leq \int_0^\infty |g(t)| \int_0^t \frac{dx}{x+t} \, dt = \ln 2 \int_0^\infty |g(t)| \, dt,$$

which completes the proof.   □

The connection between the two Hilbert transforms in (5.26) only looks completely symmetric; their interaction with $\cos xt$ and $\sin xt$ (their functioning as Hilbert transformer) differs considerably from each other; this is discussed in detail in [34]. The well-known extension of Hardy's inequality (see, e.g., [13, (7.24)])

$$\int_{\mathbb{R}} \frac{|\widehat{g}(x)|}{|x|} \, dx \lesssim \|g\|_{H^1(\mathbb{R})} \tag{5.27}$$

plays a crucial role in these considerations. The space $H^1(\mathbb{R})$ will be defined later, in the following chapter, but it is worth mentioning that here and in most situations it will mainly mean that both g and $\mathcal{H}g$ belong to $L^1(\mathbb{R})$. To get a flavor of it, recall that for $g \in L^1(\mathbb{R})$ only, $\widehat{g}$ may have an arbitrarily slow decay near infinity. Here (5.27) shows that if, in addition, $\mathcal{H}g \in L^1(\mathbb{R})$, then the Fourier transform has a prescribed decay. Since the name "Hardy's inequality" can refer to two different well-known inequalities, we dared to introduce in [34] a new name for (5.27) and its versions. Since (5.27) also involves an important aspect of the Fourier transform, we shall call it the *Fourier-Hardy inequality*, as is already mentioned in the introduction. Recall that the other inequality, the one most people think of when hearing the words "Hardy's inequality", is something like (see, e.g., [22, Theorem 330])

$$\int_0^\infty x^b \left(\frac{1}{x} R(x) \right)^p dx \leq \left(\frac{p}{|p-b-1|} \right)^p \int_0^\infty x^b \psi(x)^p \, dx, \tag{5.28}$$

where, for $\psi(x) \geq 0$ and $p > 1$, either

$$R(x) = \int_0^x \psi(t) \, dt$$

and $b < p - 1$, or

$$R(x) = \int_x^\infty \psi(t)\, dt$$

and $b > p - 1$. In addition, we mention that in this case the constant in (5.28) is sharp. There are more advanced versions, for different metrics and various weights. We shall discuss certain aspects of this issue in Chap. 7.

Let us now discuss the cases where no integrability conditions are assumed. Let f be a function with bounded variation on $\mathbb{R}$, vanishing at infinity. Since it need not be integrable, its Hilbert transform, a usual substitute for the conjugate function, may not exist. One has to use the *modified Hilbert transform* as the conjugate of a bounded function (see, e.g., [131, (3.1)] or [14, Chapter III, §1]):

$$\widetilde{\mathcal{H}}f(x) = \text{P.V.}\,\frac{1}{\pi}\int_{\mathbb{R}} f(t)\left\{\frac{1}{x-t} + \frac{t}{1+t^2}\right\} dt. \qquad (5.29)$$

This modification was introduced by Kober in [131] for L^∞ functions. Such a function is locally in any L^p; therefore it is expected to behave in regard to singularity like any L^p function. And indeed this is the case for $\widetilde{\mathcal{H}}$, since the problems near infinity are balanced by that term $\frac{t}{1+t^2}$. More precisely, the difference

$$\widetilde{\mathcal{H}}f(x) - \mathcal{H}f(x) = \int_{\mathbb{R}} f(t)\frac{t}{1+t^2}\, dt$$

is bounded for any L^p function f, $1 \le p < \infty$, by Hölder's inequality, since the function $\frac{t}{1+t^2}$ belongs to $L^{p'}$. If $f \in L^\infty$, one just has to consider separately a neighborhood of the singularity, where any L^p argument for $1 \le p < \infty$ works. In the sense of complex analysis, this is a conjugate function exactly as without that balancing summand. In fact, in most situations one can forget about the initial definition of the Hilbert transform and always use (5.29). To be precise, the kernel

$$\frac{1}{x-t} + \frac{t}{1+t^2}$$

appeared, in a sense, at least 20 years prior to Kober's publication. In many sources it is attributed to R. Nevanlinna [173] (see, e.g., [31, Appendix] or [1, Chapter III, §1]). Without details, the well-known Riesz-Herglotz kernel

$$\frac{e^{i\theta} + z}{e^{i\theta} - z},$$

where $\theta \in [-\pi, \pi]$ and z lies in the unit disk, reduces to it, up to a constant and with z in place of x, by substituting $-\cot\frac{\theta}{2} = t$ and replacing z with $\frac{z-i}{z+i}$. The latter just maps the unit disk in $\mathbb{C}$ to the upper half-plane. Correspondingly, integration over the unit circle turns to integration over $\mathbb{R}$. Also, the Riesz-Herglotz kernel becomes $\frac{t-i}{t+i}$. More precisely, in [173] the kernel never appears in the form we use. However, how to pass to it from the equivalent form it was given there is well known. This was first done in [85].

However, there is a special case where the function is not necessarily integrable but the Hilbert transform exists at *each point*. This is the already familiar Wiener algebra $W_0(\mathbb{R})$. Indeed, let $f \in W_0(\mathbb{R})$ and $f = \widehat{g}$, where $g \in L^1(\mathbb{R})$. Then the Hilbert transform is

$$
\begin{aligned}
\mathcal{H}f(x) &= \frac{1}{\pi} \int_{\to 0}^{\to \infty} \frac{f(x+t) - f(x-t)}{t} dt \\
&= \frac{2i}{\pi} \lim_{\substack{\varepsilon \to +0 \\ M \to +\infty}} \int_{-\infty}^{+\infty} g(y)e^{ixy} dy \int_{\varepsilon}^{M} \frac{\sin ty}{t} dt.
\end{aligned}
\tag{5.30}
$$

Since the absolute values of the integrals over $[\varepsilon, M]$ are bounded by an absolute constant, it is possible to pass to the limit under the integral sign, by the Lebesgue dominated convergence theorem. This yields

$$
\mathcal{H}f(x) = i \int_{-\infty}^{+\infty} g(y)e^{ixy} \operatorname{sign} y \, dy, \qquad \|\widetilde{f}\|_{W_0} = \|f\|_{W_0}.
$$

We mention that the improper integral in the definition of $\mathcal{H}f(x)$ for this class converges everywhere (and uniformly in x), but not necessarily absolutely.

We would like to mention the following relations for the cosine and sine Fourier transforms in their interrelation with the Hilbert transform:

$$
\widehat{f_s}(x) = \mathcal{H}\widehat{f_c}(x).
\tag{5.31}
$$

and

$$
\widehat{f_c}(x) = -\mathcal{H}\widehat{f_s}(x).
\tag{5.32}
$$

The formulas (5.31) and (5.32) are known, see [27, (5.42) and (5.43)] for square integrable functions. For more delicate use of these formulas in the L^1 setting, see [149] and Sect. 8.2 in Chap. 8.

Similar notions exist in the discrete setting. For the sequence $a = \{a_k\} \in \ell^1$, the discrete Hilbert transform is defined for $m \in \mathbb{Z}$ as (see, e.g., [27, (13.127)], [80])

$$
\hbar a(m) = \sum_{\substack{k=-\infty \\ k \neq m}}^{\infty} \frac{a_k}{m - k}.
\tag{5.33}
$$

If the sequence a is either even or odd, the corresponding Hilbert transforms $\hbar_e$ and $\hbar_o$ may be expressed in a special form (see, e.g., [62] or [27, (13.130) and

(13.131)]). More precisely, if a is even, with $a_0 = 0$, we have $\hbar_e(0) = 0$ and for $m = 1, 2, \ldots$

$$\hbar_e a(m) = \sum_{\substack{k=1 \\ k \neq m}}^{\infty} \frac{2ma_k}{m^2 - k^2} + \frac{a_m}{2m}. \tag{5.34}$$

If a is odd, with $a_0 = 0$, we have for $m = 0, 1, 2, \ldots$

$$\hbar_o a(m) = \sum_{\substack{k=1 \\ k \neq m}}^{\infty} \frac{2ka_k}{m^2 - k^2} - \frac{a_m}{2m}. \tag{5.35}$$

Of course, $\frac{a_0}{0}$ is considered to be zero.

In advance of a thorough study of this subject in Sect. 6.6 of Chap. 6, we mention that the discrete Hardy spaces will be denoted by h, h_e and h_o and will mean, for an ℓ^1 sequence a, that the corresponding discrete Hilbert transform is also in ℓ^1. A kind of a weak estimate for the discrete Hilbert transform has recently been obtained in [91].

5.5.1 Conditions for the Integrability of the Hilbert Transform

The above results naturally give rise to the problem of conditions which ensure the integrability of the Hilbert transform. To begin with, the following simple statement (see [49, Chapter III, 5.5 (a)] or [50, Chapter 2, §7]) holds.

Proposition 5.36 *Suppose g is a bounded function on $\mathbb{R}$ with compact support. Then $\mathcal{H}g \in L^1(\mathbb{R})$ if and only if g satisfies* (5.16).

Proof We have

$$\mathcal{H}g(x) = \frac{\widehat{g}(0)}{\pi x} + O\left(\frac{1}{x^2}\right)$$

as $|x| \to \infty$. □

For example, this simple test allows one to characterize the functions from the Schwartz class whose Hilbert transform is integrable. This happens if and only if (5.16) holds true.

On the other hand, one of the most general conditions of this kind, the so-called Zygmund-Stein condition, should be mentioned (see, e.g., [188]). It is connected to the so-called class $L \log L$, which contains all the functions satisfying

$$\int_{\mathbb{R}} |f(t)| \ln^+ |f(t)| \, dt < \infty,$$

where $\ln^+ A$ means $\ln A$ if $A > 1$ and zero otherwise. Roughly speaking, that condition in its "Zygmund part", the sufficient condition, says that a function with compact support, being in $L \log L$ and satisfying (5.16), has integrable Hilbert transform. Its necessary, "Stein side" says, with certain abuse of formality, that a function with integrable Hilbert transform is in $L \log L$ on the interval where it is like-sign.

Let us give a different condition of that sort (see [150]).

Theorem 5.37 *Suppose g is a bounded function on $\mathbb{R}$ such that for some non-negative function φ*

$$|g(t)| \leq \frac{\varphi(t)}{1 + |t|},\tag{5.38}$$

with

$$\sum_{k=0}^{\infty} k\lambda_k = \sum_{k=0}^{\infty} k \sup_{2^{k-1} < |t| \leq 2^k} \varphi(t) < \infty.\tag{5.39}$$

If it has mean zero, then $\mathcal{H}g \in L^1(\mathbb{R})$.

We postpone the proof of this test, since it will be proved within the scope of the real Hardy space in Sect. 6.2 of Chap. 6.

Roughly speaking, a bounded function with mean zero and appropriate smoothness near infinity belongs to the Hardy space, not depending on whether it is general, odd or even. A representative example of a condition that satisfies Theorem 5.37 but not Proposition 5.36 is a function that behaves as

$$g(t) = O\left(\frac{1}{t \ln^\alpha t}\right),$$

with $\alpha > 2$.

To give an idea of the line that one cannot cross, or, in other words, to show that $\alpha > 2$ is a sharp condition in the above example, we will consider an even function g, which satisfies (5.16) and whose Hilbert transform is not integrable. For a similar Pitt's example, see Example 5.21 or [130, Theorem 1(b)].

Example 5.40 Take

$$g_1(t) = \frac{1}{t \ln^2 t}$$

on $[e, \infty)$ and zero otherwise, and $g_2(t) = 1$ in $(0, 1)$ and zero otherwise. Let

$$g(t) = g_1(t) - g_2(t).$$

This function satisfies (5.16), and is integrable on $\mathbb{R}$. Consider its Hilbert transform times π for $x \in (-\infty, -2)$:

$$\int_0^1 \frac{dt}{|x|+t} + \int_e^\infty \frac{dt}{t(|x|+t)\ln^2 t}$$

$$= \frac{1}{|x|} \int_e^\infty \frac{dt}{(|x|+t)\ln^2 t} - \frac{1}{|x|} + \ln \frac{|x|+1}{|x|}$$

$$\geq \frac{1}{(|x|+e)\ln(|x|+e)} - \frac{1}{|x|} + \ln \frac{|x|+1}{|x|}.$$

Since the last two terms considered together behave as $O(\frac{1}{|x|^2})$ for large $|x|$, the Hilbert transform behaves as

$$\frac{1}{(|x|+e)\ln(|x|+e)}$$

for the considered x and thus is not integrable. What remains is to extend it to become even and take into account that the even extension possesses the same integrability properties (see again [13, Chapter III, Lemma 7.40, p. 354]).

Remark 5.41 One direction in the latter property is almost immediate. More precisely, let g be an integrable function on $\mathbb{R}_+$. If the Hilbert transform of it extended as zero to $(-\infty, 0)$, written g_0, is integrable over $\mathbb{R}$, then the Hilbert transform of its even extension g_e is integrable over $\mathbb{R}$. Indeed, by (5.34),

$$\mathcal{H}g_e(x) = \mathcal{H}_e g(x) = \frac{2}{\pi} \int_0^\infty \frac{xg(t)}{x^2-t^2} dt$$

$$= \frac{1}{\pi} \int_0^\infty \frac{g(t)}{x-t} dt + \frac{1}{\pi} \int_0^\infty \frac{g(t)}{x+t} dt. \qquad (5.42)$$

If $\mathcal{H}g_0 \in L^1(\mathbb{R})$, it remains to observe that

$$\int_{-\infty}^0 \left| \int_0^\infty \frac{g(t)}{x-t} dt \right| dx = \int_0^\infty \left| \int_0^\infty \frac{g(t)}{x+t} dt \right| dx$$

and

$$\int_0^\infty |\mathcal{H}g_e(x)| dx \leq \frac{1}{\pi} \int_0^\infty \left| \int_0^\infty \frac{g(t)}{x-t} dt \right| dx$$

$$+ \frac{1}{\pi} \int_0^\infty \left| \int_0^\infty \frac{g(t)}{x+t} dt \right| dx = \|\mathcal{H}g_0\|_{L^1(\mathbb{R})}.$$

The other direction is proved in [13, Chapter III, Lemma 7.40, p. 354] by means of atomic characterization of the real Hardy space.

It will be shown in Sect. 6.5 of the next chapter that the Hilbert transform of an odd function may behave in a completely different way. The reason that this is postponed is rather natural: too many notions and tools are used from the background of the next chapter on Hardy spaces; the Hilbert transform is touched on there since one of the ways to represent the norm in the Hardy space is via the Hilbert transform.

5.5.2 General Conditions

First of all, examining integrability of the Hilbert transform, one can test the integral over, say, $|t| \geq \frac{3}{2}|x|$. Indeed, for $x > 0$, we have

$$\int_0^\infty \left| \int_{\frac{3}{2}x}^\infty \frac{g(t)}{x-t} \, dt \right| dx$$

$$\leq \int_0^\infty |g(t)| \int_0^{\frac{2t}{3}} \frac{dx}{x-t} \, dt = \ln 3 \int_0^\infty |g(t)| \, dt.$$

The rest is estimated in a similar manner.

Further, since

$$\int_{-a}^a \frac{dt}{x-t} = 0 \tag{5.43}$$

for any $a > 0$ when the integral is understood in the principal value sense, we can always consider

$$\int_{-a}^a \frac{g(t) - g(x)}{x - t} \, dt$$

instead of the Hilbert transform truncated to $[-a, a]$.

We will prove the following result, which is less restrictive in certain respects than those above. Cf. also Sect. 8.2 of Chap. 8.

Theorem 5.44 *Let g be an integrable function on $\mathbb{R}$ which satisfies conditions (5.16),*

$$\int_{|x| \geq \frac{1}{2}} |g(x)| \log 3|x| \, dx \tag{5.45}$$

and

$$\int_{\mathbb{R}} \int_{-\frac{1}{2}\min(|x|,1)}^{\frac{1}{2}\min(|x|,1)} \left| \frac{g(x+t) - g(x)}{t} \right| dt\, dx. \tag{5.46}$$

Then $g \in H^1(\mathbb{R})$.

Since each function can be represented as the sum of its even and odd parts, we will prove Theorem 5.44 separately for odd and even functions. Thus, from now on we can consider g to be defined on $\mathbb{R}_+$ and analyze either (5.35) or (5.34) rather than the general Hilbert transform.

Odd Functions

Surprisingly, there exist many more conditions for the integrability of the Hilbert transform of an odd function than for an even one. One of the explanations is that just the odd Hilbert transform is strongly involved in the problems of the integrability of the Fourier transform (see, e.g., [141]) and has been singled out for deeper study.

Though an odd function always satisfies (5.16), not every odd integrable function belongs to $H^1(\mathbb{R})$; for counterexamples, see [164] and [146] as well as Example 5.20 above. Paley–Wiener's theorem (see [177]; for an alternative proof and discussion, see Zygmund's paper [202]) asserts that if $g \in L^1(\mathbb{R})$ is an odd and monotone decreasing function on $\mathbb{R}_+$, then $\mathcal{H}g \in L^1$, i.e., g is in $H^1(\mathbb{R})$. Recently, in [165, Theorem 6.1], this theorem has been extended to a class of functions more general than monotone ones. However, it is doubtful that these results are really practical in our situation.

Back to Theorem 5.44, we can consider

$$\frac{2}{\pi} \int_{\frac{x}{2}}^{\frac{3x}{2}} \frac{tg(t)}{x^2 - t^2}\, dt$$

instead of (5.35). Indeed, the possibility of restricting to that upper limit has been justified above. Similarly,

$$\int_0^{\infty} \left| \int_0^{\frac{x}{2}} \frac{tg(t)}{x^2 - t^2}\, dt \right| dx \leq \int_0^{\infty} |g(t)| t \int_{2t}^{\infty} \frac{dx}{x^2 - t^2}\, dt$$

$$\leq \frac{2}{3} \int_0^{\infty} |g(t)|\, dt. \tag{5.47}$$

Now, like in (5.43) and further, we have

$$\int_0^1 \left| \int_{\frac{x}{2}}^{\frac{3x}{2}} \frac{tg(t)}{x^2 - t^2}\, dt \right| dx \leq \int_0^1 \left| \int_{\frac{x}{2}}^{\frac{3x}{2}} \frac{t[g(t) - g(x)]}{x^2 - t^2}\, dt \right| dx + O\left(\int_0^{\infty} |g(t)|\, dt \right)$$

$$\leq \int_0^1 \int_{-\frac{x}{2}}^{\frac{x}{2}} \frac{|g(x+t) - g(x)|}{|t|}\, dt\, dx$$

$$+ O\left(\int_0^\infty |g(t)|\, dt\right). \tag{5.48}$$

For $x \geq 1$, we first estimate

$$\int_1^\infty \left|\int_{\frac{x}{2}}^{x-\frac{1}{2}} \frac{tg(t)}{x^2 - t^2}\, dt\right| dx \leq \int_{\frac{1}{2}}^\infty t|g(t)| \int_{t+\frac{1}{2}}^{2t} \frac{dx}{x^2 - t^2}\, dt$$

$$\leq C \int_{\frac{1}{2}}^\infty |g(t)| \ln 3t\, dt$$

and

$$\int_1^\infty \left|\int_{x+\frac{1}{2}}^{\frac{3x}{2}} \frac{tg(t)}{x^2 - t^2}\, dt\right| dx \leq \int_{\frac{3}{2}}^\infty t|g(t)| \int_{\frac{2t}{3}}^{t-\frac{1}{2}} \frac{dx}{t^2 - x^2}\, dt$$

$$\leq C \int_{\frac{1}{2}}^\infty |g(t)| \ln 3t\, dt.$$

These two bounds lead to the logarithmic condition (5.45). The remaining integral

$$\int_1^\infty \left|\int_{x+\frac{1}{2}}^{x-\frac{1}{2}} \frac{tg(t)}{x^2 - t^2}\, dt\right| dx$$

is estimated exactly like that in (5.48). Applying (5.43), we obtain

$$\int_1^\infty \left|\int_{x-\frac{1}{2}}^{x+\frac{1}{2}} \frac{tg(t)}{x^2 - t^2}\, dt\right| dx \leq \int_1^\infty \left|\int_{x-\frac{1}{2}}^{x+\frac{1}{2}} \frac{t[g(t) - g(x)]}{x^2 - t^2}\, dt\right| dx$$

$$+ O\left(\int_0^\infty |g(t)|\, dt\right)$$

$$\leq \int_1^\infty \int_{-\frac{1}{2}}^{\frac{1}{2}} \frac{|g(x+t) - g(x)|}{|t|}\, dt\, dx$$

$$+ O\left(\int_0^\infty |g(t)|\, dt\right). \tag{5.49}$$

Combining all the obtained estimates, we arrive at the required result. $\qquad\square$

Even Functions

While an odd function always satisfies (5.16), in the case of even functions the situation is more delicate: the function must have the cancelation property already on the half-axis:

$$\int_0^\infty g(t)\,dt = 0. \tag{5.50}$$

With this in hand, the proof goes along the same lines as that for odd functions. The only problem is that an estimate like (5.47) does not follow immediately from formula (5.34). However, using the above remark on the cancelation property for g on the half-axis, we can rewrite (5.34) as

$$\mathcal{H}_e g(x) = \frac{2}{\pi} \int_0^\infty g(t)\left[\frac{x}{x^2 - t^2} - \frac{2}{x}\right]dt. \tag{5.51}$$

Now,

$$\int_0^\infty \left|\int_0^{\frac{x}{2}} \frac{t^2 g(t)}{x(x^2 - t^2)}\,dt\right| dx \le \int_0^\infty |g(t)|t^2 \int_{2t}^\infty \frac{dx}{x(x^2 - t^2)}\,dt$$

$$\le \frac{1}{6}\int_0^\infty |g(t)|\,dt. \tag{5.52}$$

This additional term $\frac{2}{x}$ does not affect the other estimates of the previous subsection. The proof is complete.   □

5.6 Summability to the Hilbert Transform

Of course, the two main operators of the one-dimensional harmonic analysis, the Fourier transform and the Hilbert transform, have different "obligations" but certain general features are similar. For example, the idea of summability so natural for the Fourier transform makes sense for the Hilbert transform as well. The results in this section are about Cesàro's and more general summability of the conjugate function (5.11)

$$\frac{1}{\pi} \int_0^\infty \int_{-\infty}^\infty f(t) \sin(t - x)u\,dt\,du. \tag{5.53}$$

In his review of Wiener's book "The Fourier Integral and Certain of its Applications" [56], Hille writes: *"An analyst of the older generation would probably understand the term Fourier integral to refer to the formula"* (5.53), *"which is valid for a very restricted class of functions"*. This is one of the forms of conjugation; it is used, for example, by Titchmarsh in [54] while showing, in a formal way, how the Hilbert transform appears as a natural conjugation in the non-periodic case.

The following result (see [57, Vol.II, Chapter XVI, Theorem 1.22]) demonstrates how Cesàro's (or $(C, 1)$ in different notation) method works in this setting reducing the conjugation to the Hilbert transform.

Theorem 5.54 *If* $\dfrac{|f(t)|}{1 + |t|}$ *is integrable on* $\mathbb{R}$, *then the* $(C, 1)$ *means*

$$-\frac{1}{\pi} \int_{-\infty}^{\infty} f(x + t) \left[\frac{1}{t} - \frac{\sin Nt}{Nt^2} \right] dt$$

converge to the Hilbert transform $\mathcal{H}f(x)$ *almost everywhere as* $N \to \infty$.

The λ-summability on the half-axis means that we have to deal with (estimate) are naturally defined as

$$\lim_{N \to \infty} \int_0^\infty \lambda\left(\frac{u}{N}\right) \int_{-\infty}^\infty f(x + t) \sin ut \, dt \, du. \tag{5.55}$$

Indeed, (5.53) resembles the Fourier inverse (for the conjugate function), or, in other words, the inner integral in (5.53) resembles the Fourier transform times modulation. By this, standard summability comes into play by introducing the multiplier $\lambda\left(\frac{u}{N}\right)$, with varying parameter N; the simplest example is $\lambda\left(\frac{u}{N}\right) = 1 - \frac{u}{N}$ for $0 \le u \le N$, and zero otherwise.

In Proposition 8.2.7 of [10, Section 8.2.3], related results are expressed in terms of the Hilbert transform of λ. We study the convergence of (5.55) in terms of the Fourier transforms of λ and estimates of such Fourier transforms. The conditions we are going to impose on λ slightly resemble the conditions posed on the functions generating summability methods in [197, Section 5.3], while studying the convergence of the appropriate multidimensional means to the Riesz transforms.

A natural set where the almost everywhere convergence of the means for an integrable function f is usually studied is the set of its Lebesgue points; for the definition, see (2.9). It will be convenient for us to write the defining relation of a Lebesgue point x of f in the equivalent form

$$\frac{1}{\delta} \int_0^\delta |f(x + u) - f(x - u)| \, du = o(1)$$

as $\delta \to 0+$.

We will prove a general result, where the summability is defined by (5.55). In the formulation of the theorem as well as in its proof, the cosine and sine Fourier transforms of λ and λ' will be involved (for the definitions, see (4.5) and (4.6) in the previous chapter. For example, $\widehat{\lambda'_c}$ means the cosine Fourier transform of λ'.

Theorem 5.56 *Let $f \in L^1(\mathbb{R})$. Let λ be an integrable and locally absolutely continuous function on $(0, \infty)$, such that*

$$\lim_{t \to \infty} \lambda(t) = 0$$

and

$$\lim_{t \to 0+} \lambda(t) = 1,$$

the derivative λ' be integrable on $(0, \infty)$ and

$$|\widehat{\lambda'_c}(t)| \le \varphi(t),$$

where φ is locally absolutely continuous on $(1, \infty)$, uniformly bounded on $[1, \infty)$, and satisfies

$$\int_1^\infty \frac{\varphi(t)}{t} \, dt < \infty \tag{5.57}$$

and

$$\int_1^\infty |\varphi'(t)| \, dt < \infty. \tag{5.58}$$

Then (5.53) is λ-summable by means of (5.55) to $-\mathcal{H}f(x)$ at every x, where it exists, and which is a Lebesgue point of f, and so almost everywhere.

Proof Since both f and λ are Lebesgue integrable, we can change the order of integration in (5.55):

$$\lim_{N \to \infty} \int_0^\infty \lambda\left(\frac{u}{N}\right) \int_{-\infty}^\infty f(x+t) \sin ut \, dt \, du$$

$$= \lim_{N \to \infty} \int_0^\infty [f(x+t) - f(x-t)] \int_0^N \lambda\left(\frac{u}{N}\right) \sin ut \, du \, dt$$

$$= \lim_{N \to \infty} \int_0^\infty [f(x+t) - f(x-t)] \, N\widehat{\lambda_s}(Nt) \, dt.$$

Integrating by parts, we obtain

$$N\widehat{\lambda_s}(Nt) = \frac{1}{t} + \frac{1}{t}\widehat{\lambda'_c}(Nt) \tag{5.59}$$

provided that

(i) λ is locally absolutely continuous on $(0, \infty)$,

(ii) $\lim\limits_{t \to \infty} \lambda(t) = 0$,

and

(iii) $\lim\limits_{t \to 0+} \lambda(t) = 1$

hold.

For any $\delta > 0$, we get

$$\lim_{N \to \infty} \int_\delta^\infty [f(x+t) - f(x-t)] N\widehat{\lambda_s}(Nt)\, dt$$
$$= \int_\delta^\infty \frac{f(x+t) - f(x-t)}{t}\, dt$$

provided that

(iv) $\lim\limits_{s \to \infty} \widehat{\lambda_c'}(s) = 0$

holds. Furthermore,

$$\int_{\frac{1}{N}}^\delta [f(x+t) - f(x-t)] N\widehat{\lambda_s}(Nt)\, dt$$
$$= \int_{\frac{1}{N}}^\delta \frac{f(x+t) - f(x-t)}{t}\, dt$$
$$+ \int_{\frac{1}{N}}^\delta [f(x+t) - f(x-t)] \frac{1}{t}\widehat{\lambda_c'}(Nt)\, dt. \tag{5.60}$$

For the last term on the right-hand side of (5.60), the assumption of the theorem implies

$$\int_{\frac{1}{N}}^\delta |f(x+t) - f(x-t)| \frac{1}{t}|\widehat{\lambda_c'}(Nt)|\, dt$$
$$\leq \int_{\frac{1}{N}}^\delta |f(x+t) - f(x-t)| \frac{\varphi(Nt)}{t}\, dt. \tag{5.61}$$

Integrating by parts, we obtain

$$\int_{\frac{1}{N}}^\delta |f(x+t) - f(x-t)| \frac{\varphi(Nt)}{t}\, dt$$
$$= \left[\frac{1}{t}\int_0^t |f(x+u) - f(x-u)|\, du\, \varphi(Nt) \right]_{\frac{1}{N}}^\delta$$

$$+ \int_{\frac{1}{N}}^{\delta} \frac{1}{t} \int_0^t |f(x+u) - f(x-u)| \, du \, \frac{\varphi(Nt)}{t} \, dt$$

$$- \int_{\frac{1}{N}}^{\delta} \frac{1}{t} \int_0^t |f(x+u) - f(x-u)| \, du \, N\varphi'(Nt) \, dt. \tag{5.62}$$

The integrated terms are $o(1)$ times $\varphi(N\delta)$ and $\varphi(1)$, respectively. Hence both integrated terms are $o(1)$ provided x is a Lebesgue point, since φ is uniformly bounded on $[1, \infty)$.

If x is a Lebesgue point, then for the last two integrals on the right-hand side of (5.62) to be $o(1)$, it suffices that

$$\int_{\frac{1}{N}}^{\delta} \frac{\varphi(Nt)}{t} \, dt = \int_1^{N\delta} \frac{\varphi(t)}{t} \, dt$$

and

$$\int_{\frac{1}{N}}^{\delta} N|\varphi'(Nt)| \, dt = \int_1^{N\delta} |\varphi'(t)| \, dt$$

to be uniformly bounded. This is the case if

(v) $\int_1^\infty \frac{\varphi(t)}{t} \, dt < \infty$

and

(vi) $\int_1^\infty |\varphi'(t)| \, dt < \infty$,

respectively, hold true.

Finally,

$$\left| \int_0^{\frac{1}{N}} [f(x+t) - f(x-t)] \, N\widehat{\lambda_s}(Nt) \, dt \right|$$

$$\lesssim N \int_0^{\frac{1}{N}} |f(x+t) - f(x-t)| \, dt = o(1),$$

provided that

(vii) $|\widehat{\lambda_s}|$ is bounded,

since x is a Lebesgue point.

We now have to check that (i)–(vii) are valid under assumptions of the theorem. Indeed, (i)–(iii) are assumed. Since the Fourier transform of an integrable function vanishes at infinity, by the Riemann–Lebesgue lemma, (iv) holds. Similarly, the Fourier transform of an integrable function is bounded, which implies (vii). Finally, (v) and (vi) are just (5.57) and (5.58), respectively, which completes the proof. $\square$

Back to Titchmarsh's book, the result of [54, Theorem 107] is given for any (C, α) method. In turn, Theorem 5.54 follows from it by taking $\alpha = 1$ and integrating then by parts in the integral representation of $\widehat{\lambda_s}(Nt)$. In fact, [54, Theorem 107] is a corollary of the latter theorem.

Corollary 5.63 *Let $f \in L^1(\mathbb{R})$. Then, for any positive α, the integral (5.53) is summable (C, α) to $-\mathcal{H}f(x)$ at every x, where it exists, and which is a Lebesgue point of f, and so almost everywhere.*

Proof First, it suffices to suppose that $0 < \alpha < 1$. The (C, α) summability means that we estimate

$$\lim_{N \to \infty} \int_0^N \left(1 - \frac{u}{N}\right)^\alpha \int_{-\infty}^\infty f(x + t) \sin ut \, dt \, du$$

$$= \lim_{N \to \infty} \int_0^\infty [f(x + t) - f(x - t)] \int_0^N \left(1 - \frac{u}{N}\right)^\alpha \sin ut \, du \, dt.$$

We have to check the assumptions of Theorem 5.56 for $\lambda(t) = (1 - t)_+^\alpha$, where the standard subindex $+$ means that λ takes the indicated value only when non-negative, and vanishes if negative. All the conditions on λ are readily checked. Since the Fourier transform of its derivative is $O(t^{-\alpha})$, we can take $\varphi(t) = t^{-\alpha}$. It remains to observe that (5.57) and (5.58) reduce to the integrability of $t^{-1-\alpha}$ over $(1, \infty)$. □

The proof of Theorem 5.56 shows that the assumptions of the theorem are not very restrictive and seem quite natural. However, it is worth discussing certain options for these assumptions and reasonable substitutions.

By the Riemann–Lebesgue lemma, (iv) holds, for example, if λ' is integrable. This is the case if, say, λ is of bounded variation. One more option for this is assuming λ to be general monotone. Additionally assuming then

$$\int_0^\infty \frac{|\lambda(t)|}{t} \, dt < \infty$$

leads to the boundedness of total variation; for all these notions and facts, see, e.g., [165]. However, since we assume λ to be integrable, a natural way to have λ' integrable is to assume λ belong to the Sobolev space $W^{1,1}$.

It is interesting that not proceeding from (5.60) to (5.61) and then to (5.62) but integrating by parts directly in (5.60), which seems to be more precise, leads to worse results and conditions. For example, Corollary 5.63 is not verifiable then.

There is an interesting way to guarantee (5.57), or more precisely, the same integral with $|\widehat{\lambda_c'}(t)|$ in place of $\varphi(t)$. In this case, the finiteness of the integral can be ensured by λ' belong to the *local Hardy space*. For many details about this space, see [113] and a recent paper [90].

In conclusion, there are many interesting combinations to substitute for the assumptions of Theorem 5.56. Also, [178] can be mentioned as a paper where a similar problem is studied.

Chapter 6
Hardy Spaces and their Subspaces

As we have previously stated on a number of occasions, if g is integrable, its Hilbert transform is not necessarily integrable. Moreover, it can be even not locally integrable. When the Hilbert transform is integrable, we say that g is in *the (real) Hardy space* $H^1 := H^1(\mathbb{R})$. There are a variety of its characterizations (or, equivalently, definitions). The one given by means of the Hilbert transform is both important and convenient.

Definition 6.1 The real Hardy space $H^1 := H^1(\mathbb{R})$ is the subspace of $L^1(\mathbb{R})$ which consists of functions with integrable Hilbert transform. It is endowed with the norm

$$\|g\|_{H^1(\mathbb{R})} = \|g\|_{L^1(\mathbb{R})} + \|\mathcal{H}g\|_{L^1(\mathbb{R})}. \tag{6.2}$$

A natural question is whether only $H^1(\mathbb{R})$ will be considered in what follows. The negative answer is two-fold. First, though indeed $H^1(\mathbb{R})$ will be our main object in this chapter, it will be shown that certain of its subspaces and counterparts are of at least equal interest and importance. On the other hand, we shall discuss or mention other $H^p(\mathbb{R})$ spaces, at least in a passing manner. For instance, a celebrated book by Koosis [30] studies, from various points of view, the H^p spaces for all $p > 0$. Let us specify our interest in various aspects of the theory of Hardy spaces. Very little attention will be paid to the case $p = \infty$, mostly for functions not satisfying any integrability condition rather than belonging just to L^∞. There will be a brief discussion of the case $1 < p < \infty$ along with an explanation that the equivalence of H^p spaces with these p to the L^p spaces does not suggest further options for their scrutiny. On the other hand, a rich theory of H^p spaces, with $0 < p < 1$, will be touched on in a minimal way only because of the lack of the author's personal involvement in the problems concerning these spaces, at least till recently. What will be mentioned is that their atomic theory is very close to that of the H^1 space.

© The Author(s), under exclusive license to Springer Nature Switzerland AG 2021
E. Liflyand, *Harmonic Analysis on the Real Line*, Pathways in Mathematics,
https://doi.org/10.1007/978-3-030-81892-0_6

6.1   Some Starting Points

If $g \in H^1(\mathbb{R})$, then, as mentioned above, (5.16) holds true. The functions $g \in L^1(\mathbb{R})$, which satisfy (5.16), are sometimes called *wavelet functions* (see, e.g., [183]), and the corresponding space is denoted by $L_0^1(\mathbb{R})$. By this, strictly speaking, $H^1(\mathbb{R})$ is a subspace of $L_0^1(\mathbb{R})$, which, in turn, is a subspace of $L^1(\mathbb{R})$. For example, $H^1(\mathbb{R})$ is dense in $L_0^1(\mathbb{R})$ but not in the whole $L^1(\mathbb{R})$.

An odd function always satisfies (5.16). However, not every odd integrable function belongs to $H^1(\mathbb{R})$; for counterexamples, see [164] (or Example 5.20) and [146].

For functions in $L^p(\mathbb{R})$, $1 < p < \infty$, the situation is completely different. A deep theorem due to M. Riesz asserts that H^p spaces for these p, in fact, coincide with L^p. One of the ways to prove it is by interpolation, possible because of the weak estimate in Theorem 5.12 and the strong L^2 estimate (5.10). Correspondingly, the existence almost everywhere of the Hilbert transform of a function in $L^p(\mathbb{R})$ can be proved as in the previous chapter for integrable functions. More precisely, we have

$$\|\mathcal{H}g\|_{L^p(\mathbb{R})} \leq C_p \|g\|_{L^p(\mathbb{R})}, \tag{6.3}$$

which is often called the *M. Riesz projection theorem*. Though we have only given a hint how M. Riesz's theorem can be proved, we shall use it quite a few times. Here C_p is the well-known Pichorides (Gohberg-Krupnik for $p = 2^k$, see, e.g., [127]) constant:

$$C_p = \begin{cases} \tan \frac{\pi}{2p}, & 1 < p \leq 2, \\[2mm] \cot \frac{\pi}{2p}, & 2 \leq p < \infty. \end{cases}$$

It is clear that the constant C_p depends on p in such a way that $\lim_{p \to 1} C_p = \infty$. The same, of course, occurs as $p \to \infty$. This is also supported by Example 5.3 in the previous chapter.

One more issue of outstanding importance is that the dual space of H^1 is the so-called BMO space. It was a "revolution" in this theory in the beginning of the 70-s when the celebrated paper by Ch. Fefferman and Stein [99] appeared. In fact, the statements were published even earlier, in [97], in parallel with several other prominent publications of Ch. Fefferman over that period. The subject of this duality and its applications will not be touched on in these notes.

If one deals with functions on the half-axis, there are three main options to stay within the framework of the general theory of Hardy spaces: to consider only odd or only even functions, or to extend the given function by zero. We have seen that in the context of the real Hardy space the two last options coincide, in a sense (cf. Example 5.40).

Definition 6.4 If the odd Hilbert transform is integrable, we shall denote the corresponding Hardy space by $H_o^1(\mathbb{R}_+)$, or sometimes simply H_o^1. Symmetrically, if the even Hilbert transform is integrable, we shall denote the corresponding Hardy space by $H_e^1(\mathbb{R}_+)$, or sometimes simply H_e^1.

It is obvious that to have (5.16) for $H_e^1(\mathbb{R}_+)$ reduces to (5.50).

6.2 Atomic Characterization

It turns out that for the study of various properties of $H^1(\mathbb{R})$ and its subspaces, Definitions 6.1 and 6.4 are not enough or sometimes extremely inconvenient. We will need some of the other characterizations of the real Hardy space, and this is the subject of this section and the following one. More precisely, the real Hardy space can be characterized in different ways than (6.2). Maybe the one via maximal function can be considered as the most comprehensive (or, so to say, catch-all) but we skip it and rather concentrate on that via the so-called *atomic decomposition*.

6.2.1 Atoms

Let $a(x)$ denote an atom (a $(1, q, 0)$-atom), $1 < q \leq \infty$, a function of compact support, to wit

$$\operatorname{supp} a \subset I = [c, d], \qquad -\infty < c < d < +\infty; \tag{6.5}$$

which satisfies the *size condition* (L^q normalization), with $|I| = d - c$,

$$\|a\|_q \leq \frac{1}{|I|^{1-\frac{1}{q}}}; \tag{6.6}$$

and the *cancelation condition*

$$\int_{\mathbb{R}} a(x) \, dx = 0. \tag{6.7}$$

It is well-known (see, e.g., [49, Chapter III, 2.2]; also [13] or [50]) that

$$\|f\|_{H^1(\mathbb{R})} \sim \inf \left\{ \sum_k |\lambda_k| < \infty : f(x) = \sum_k \lambda_k a_k(x) \right\}, \tag{6.8}$$

where a_k are the above described atoms, with certain q fixed, and $\sum_k |\lambda_k| < \infty$ ensures that the sum $\sum_k \lambda_k a_k(x)$ converges in the L^1 norm. Any fixed $q > 1$ leads to the same space, which is equivalent to the real Hardy space defined above via the Hilbert transform. Taking for each k its q_k leads to new, wider spaces provided $q_k \to 1$ (see [189] and discussion below) or, under some additional assumptions, to the atomic characterization of $L^1(\mathbb{R})$ (see [58] and [59]).

Since today this characterization is discussed in detail in many sources, we rather dwell on some specific applications. An obvious advantage of atomic characterization is that in many situations it suffices to verify a certain statement only on atoms. However, this is not a "law of Nature" and one should be cautious in using this approach; see a celebrated example due to Bownik in [79].

What is of crucial importance for us is that all the characterizations of the real Hardy space are equivalent. This means, in particular, that proving a statement by means of atomic decomposition we have the same fact in terms of (6.2). We shall not consider the details in full; let us prove that the L^1 norm of the Hilbert transform of a $(1, q, 0)$-atom is bounded by a constant C depending only on q. More precisely, we have

Proposition 6.9 *There holds*

$$\int_{\mathbb{R}} |\mathcal{H}a(x)|\, dx \leq \frac{C}{q-1}. \tag{6.10}$$

Proof Denoting $r = \frac{c+d}{2}$, we split the integral in (6.10) into two: over $|x - r| > 2|I|$ and over $|x - r| \leq 2|I|$. By (6.7), the first one can be rewritten as

$$\pi \int_{|x-r|>2|I|} |\mathcal{H}a(x)|\, dx = \int_{|x-r|>2|I|} \left| \int_c^d a(t) \left[\frac{1}{x-t} - \frac{1}{x-r} \right] dt \right| dx.$$

The right-hand side is estimated by

$$\int_{|x-r|>2|I|} \frac{1}{|x-r|^2} \int_c^d |a(t)|\, |t-r|\, dt\, dx.$$

Applying Hölder's inequality and (6.6), we obtain the bound, times a constant multiple,

$$\int_{|x-r|>2|I|} \frac{1}{|x-r|^2} |I|^{\frac{q'+1}{q'}} \frac{1}{|I|^{1-\frac{1}{q}}} \leq 2.$$

For the integral over $|x - r| \leq 2|I|$, we apply Hölder's inequality immediately:

$$\int_{|x-r|\leq 2|I|} |\mathcal{H}a(x)|\, dx \leq \left(\int_{|x-r|\leq 2|I|} dx \right)^{1-\frac{1}{q}} \left(\int_{\mathbb{R}} |\mathcal{H}a(x)|^q\, dx \right)^{\frac{1}{q}}.$$

By the M. Riesz theorem, the right-hand side is controlled with

$$\left(\int_{|x-r|\leq 2|I|} dx \right)^{1-\frac{1}{q}} \left(\int_{\mathbb{R}} |a(x)|^q\, dx \right)^{\frac{1}{q}} \leq 1.$$

Combining the obtained estimates, we complete the proof.   □

Of course, we could make the calculations slightly better, observing like in, say, [50, Chapter 2, §5], that the Hilbert transform of $A_s(t) := s\, a(st)$ is $s\, \mathcal{H}a(sx)$ and A_s is also an atom with support $\frac{|I|}{s}$, on the one hand, and that translation preserves atoms as well as the sizes of their supports.

A natural question is: how $\frac{1}{q-1}$ appears on the right-hand side of (6.10)? The answer is: from the M. Riesz theorem; see (6.3) and below. Writing "controlled" in the above proof means times this factor multiple.

6.2.2   Atomic Proof of the Fourier-Hardy Inequality

Despite the above warning, in most situations atomic characterization works pretty well. To illustrate this, let us make use of it for proving (5.27). It will be proved if we show that for any atom a there holds

$$\int_{\mathbb{R}} \frac{|\hat{a}(x)|}{|x|}\, dx \leq C, \tag{6.11}$$

with C independent of the atom. Let a be a $(1, q, 0)$-atom, $1 < q \leq 2$, and let us split the integral on the left-hand side into two: over $|x| \leq \frac{1}{|I|}$ and over $|x| > \frac{1}{|I|}$. For the first one, (6.7), Hölder's inequality and (6.6) yield

$$\int_{|x|\leq\frac{1}{|I|}} \frac{|\hat{a}(x)|}{|x|}\, dx = \int_{|x|\leq\frac{1}{|I|}} \frac{1}{|x|} \left| \int_c^d a(t)[e^{-ixt} - e^{-ixa}]\, dt \right| dx$$

$$\leq \int_{|x|\leq\frac{1}{|I|}} \frac{1}{|x|} \|a\|_{L^q} \left(\int_c^d |e^{-ixt} - e^{-ixa}|^{q'}\, dt \right)^{\frac{1}{q'}} dx$$

$$\leq \int_{|x|\leq\frac{1}{|I|}} |I|^{\frac{1}{p}-1} \left(\int_c^d (t-a)^{q'}\, dt \right)^{\frac{1}{q'}} dx$$

$$\leq \frac{2}{(q'+1)^{\frac{1}{q'}}}.$$

For the second integral, applying Hölder's inequality and the Hausdorff-Young inequality with any rough absolute constant C (cf. the discussion after (4.10) in Chap. 4), we obtain

$$\int_{|x|>\frac{1}{|T|}} \frac{|\widehat{a}(x)|}{|x|} dx \leq \left(\int_{|x|>\frac{1}{|T|}} |x|^{-q} dx\right)^{\frac{1}{q}} \left(\int_{\mathbb{R}} |\widehat{a}(x)|^{q'} dx\right)^{\frac{1}{q'}}$$

$$\leq \frac{C}{(q-1)^{\frac{1}{q}}} (d-c)^{\frac{1}{p}-1} \left|\int_c^d |a(t)|^q dt\right|^{\frac{1}{q}}.$$

Using (6.6) for the last integral, we arrive at the required bound.

6.2.3 A Postponed Proof

One more illustration comes from the postponed

Proof of Theorem 5.37 The scheme of the proof is the same as in [50, Chapter 2, §7] for the case where $g(t) = O(\frac{1}{1+t^2})$. We denote $g_0(t) = g(t)$ when $|t| \leq 1$ and $g_0(t) = 0$ otherwise. Further, for $k = 1, 2, \ldots$, define

$$g_k(t) = \begin{cases} g(t), & 2^{k-1} < |t| \leq 2^k, \\ 0, & \text{otherwise} \end{cases}$$

and

$$c_k = \int_{\mathbb{R}} g_k(t) \, dt = \int_{2^{k-1} < |t| \leq 2^k} g(t) \, dt.$$

We observe that under the assumptions of the theorem g is an integrable function, since

$$|c_k| \leq \int_{2^{k-1} < |t| \leq 2^k} \frac{\lambda_k}{t} \, dt. \tag{6.12}$$

We have

$$g(t) = \sum_{k=0}^{\infty} g_k(t).$$

Let us also denote

$$S_k = \sum_{j \geq k} c_j;$$

it follows from the assumption of the theorem that $S_0 = 0$.

Taking a bounded function η supported in $\{t : |t| \leq 1\}$ and satisfying the equality

$$\int_{\mathbb{R}} \eta(t)\, dt = 1,$$

and denoting $\eta_k(t) = \frac{1}{2^k} \eta(\frac{t}{2^k})$, we obtain

$$\int_{\mathbb{R}} \eta_k(t)\, dt = 1$$

and

$$g(t) = \sum_{k=0}^{\infty} [g_k(t) - c_k \eta_k(t)] + \sum_{k=0}^{\infty} c_k \eta_k(t). \qquad (6.13)$$

The first sum is

$$\sum_{k=0}^{\infty} [g_k(t) - c_k \eta_k(t)] = \sum_{k=0}^{\infty} \frac{1}{2^k} \int_{2^{k-1} < |x| \leq 2^k} [g_k(t) - g(x)\eta(\frac{t}{2^k})]\, dx$$

$$= \sum_{k=0}^{\infty} B_k(t).$$

We have

$$\int_{\mathbb{R}} B_k(t)\, dt = \int_{2^{k-1} < |x| \leq 2^k} g(x)\, dx \left[1 - \int_{\mathbb{R}} \eta_k(t)\, dt \right] = 0.$$

The function $B_k(t)$ is supported in $\{t : |t| \leq 2^k\}$. Finally, for $2^{k-1} < |t| \leq 2^k$, there holds $|g_k(t)| \leq C\frac{\lambda_k}{2^k}$. Combining this with (6.12), we obtain the estimate

$$\|B_k(t)\|_\infty \leq C\frac{\lambda_k}{2^k}.$$

By this, $B_k(t) = C\lambda_k A_k(t)$, where $A_k(t)$ is an atom supported in $\{t : |t| \leq 2^k\}$. Here, the numbers λ_k are the coefficients of an atomic decomposition of g, and C is an absolute constant.

Concerning the second sum in (6.13), it can be represented in an equivalent form due to the mean zero property ($S_0 = 0$):

$$\sum_{k=1}^{\infty} S_k (\eta_k - \eta_{k-1}).$$

Each value $\eta_k(t) - \eta_{k-1}(t)$ is a multiple of an analogous atom. The numbers S_k are appropriate for an atomic decomposition, since

$$\sum_{j=0}^{\infty} |S_j| = \sum_{j=0}^{\infty} \left| \sum_{k=j}^{\infty} c_k \right| \leq \sum_{j=0}^{\infty} \sum_{k=j}^{\infty} \int_{2^{k-1} < |t| \leq 2^k} |g(t)| \, dt$$

$$\leq C \sum_{j=0}^{\infty} \sum_{k=j}^{\infty} \lambda_k = C \sum_{k=1}^{\infty} k \lambda_k,$$

and (5.39) completes the proof. $\square$

6.2.4 More About Atomic Characterization

This is apparently the proper moment to mention that atomic characterization is an excellent tool for treating H^p spaces, with $0 < p < 1$. Like H^1 is a smooth-acting part of L^1 in a variety of problems, H^p is such as opposed to L^p, with $0 < p < 1$, even to greater extent because of the pathological nature of these L^p (for instance, non-trivial functionals do not exist in these L^p). For this, $(p, q, 0)$-atoms are used, with a proper re-normalization in (6.6); we omit the details. What we should mention is that such atoms are not always optimal. For this reason, special atoms were constructed by Miyachi in, e.g., [171]. They were effectively applied for Hausdorff operators on H^p spaces, with $0 < p < 1$, in [158].

A different problem atomic in nature is considered by Sweezy in [189] (see also [59]). The problem was what happens between the two classical cases L^1, with L^∞ dual, and H^1, with BMO dual. To solve this, modifications of the usual $(1, q, 0)$-atoms were introduced. They were just divided by $(q')^r$, and the series of expansions in these modified atoms as $q \to 1$ form a scale of nested spaces between H^1 and L^1 according to r when it runs from 1 to infinity. Correspondingly, their dual spaces form a scale from BMO to L^∞. It is worth mentioning that atomic description of these spaces is the only known way to define them.

More discussion on the cancelation condition (6.7) is in order. It is written in [49, Chapter III, 5.6], that dropping it, that is, dealing with all the "atoms" a_k satisfying

(6.5) and (6.6), with $q = \infty$ for certainty, but not necessarily to (6.7) (in [50] they are called the "faux" atoms) leads to the functions f that are obtainable as sums

$$f(x) = \sum_k \lambda_k a_k(x),$$

which form exactly all $L^1(\mathbb{R})$. Indeed, given $f \in L^1(\mathbb{R})$, for each k, we define

$$f_k(x) = 2^k \int_{\frac{l}{2^k}}^{\frac{l+1}{2^k}} f(t)\,dt, \quad x \in \left(\frac{l}{2^k}, \frac{l+1}{2^k}\right).$$

The functions f_k form the "dyadic martingale" sequence associated to f. Clearly, $f_k \to f$ in L^1 norm, as $k \to \infty$. Indeed, this is immediate for step functions that, in turn, are dense in L^1. It follows from this that there exist $k_j \to \infty$ for which

$$\sum_j \|f_{k_j} - f_{k_{j-1}}\|_{L^1(\mathbb{R})} < \infty.$$

We can now write

$$f = f_{k_0} + \sum_{j=1}^{\infty} \left(f_{k_j} - f_{k_{j-1}}\right).$$

Since $f_{k_j} - f_{k_{j-1}}$ is constant on each dyadic interval of length $\frac{1}{2^{k_j}}$, the resulting sum gives a kind of atomic decomposition of f, whose putative atoms fail to have the cancelation property.

For a different atomic type characterization of L^1, see the already mentioned [58].

One can consider an intermediate situation where the cancelation property is assumed only for atoms with small supports, or, in other words, for atoms associated with the intervals longer than 2 (in several dimensions the balls of radius ≥ 1) no cancelation property is required. This leads to the so-called local Hardy space; see [113], related is also a recent paper [90].

6.3 Molecular Characterization

The idea of molecular decompositions is due to Coifman and Weiss [87]; a good description can be found in [13, Chapter III, §7]. Recall that a function M is called a *molecule* (centered at $x_0 \in \mathbb{R}$) if it satisfies (5.16) and

$$\left(\int_{\mathbb{R}} |M(x)|^2 dx\right)^{\frac{1}{4}} \left(\int_{\mathbb{R}} |x - x_0|^2 |M(x)|^2 dx\right)^{\frac{1}{4}} < \infty.$$

The left-hand side is called a molecular norm of M and is denoted by $N(M)$. Every $(1, 2, 0)$ atom $a(t)$ is a molecule. Indeed,

$$\int_c^d |a(t)|^2 (t - c)^2 dt \leq (d - c),$$

and together with (6.6), with $p = 2$, this implies the finiteness of the molecular norm.

A handy machinery is delivered by the following fact (see, e.g., [13, Chapter III, §7, p.328]).

Proposition 6.14 *A function g belongs to* $H^1(\mathbb{R})$ *if and only if*

$$g(x) = \sum_{j=1}^{\infty} M_j(x)$$

for almost all x, with the M_j*-s being molecules which satisfy*

$$\sum_{j=1}^{\infty} N(M_j) < \infty,$$

where

$$N(M) = \|M\|_{L^2(\mathbb{R})}^{\frac{1}{2}} \|t M(t)\|_{L^2(\mathbb{R})}^{\frac{1}{2}}.$$

To get an idea of how molecules work, let us prove that such an M without satisfying (5.16) is Lebesgue integrable. Adding (5.16) implies that it belongs to $H^1(\mathbb{R})$. We follow [13, Chapter III, §7].

The former is very simple ([13, Chapter III, §7, Lemma 7.11]). Indeed, applying the Cauchy-Bunyakovskii-Schwarz inequality, we get

$$\int_{\mathbb{R}} |M(t)| \, dt = \left(\int_{|t| \leq r} + \int_{|t| > r} \right) |M(t)| \, dt$$

$$\leq \sqrt{2r} \, \|M\|_{L^2(\mathbb{R})} + \sqrt{\frac{2}{r}} \, \|t M(t)\|_{L^2(\mathbb{R})}.$$

Both norms are finite by definition of molecule. One can take $r = 1$ but also one can play with r in certain situations. The latter is more important, since taking

$$r = \frac{\|t M(t)\|_{L^2(\mathbb{R})}}{\|M\|_{L^2(\mathbb{R})}},$$

we obtain

$$\|M\|_{L^1(\mathbb{R})} \le 2^{\frac{3}{2}} \|M\|_{L^2(\mathbb{R})}^{\frac{1}{2}} \|tM(t)\|_{L^2(\mathbb{R})}^{\frac{1}{2}},$$

which gives rise to molecular characterization of $H^1(\mathbb{R})$.

Further, let (5.16) hold. Since we already know that $M \in L^1(\mathbb{R})$, using the equivalent Definition 6.1 for H^1, it suffices to prove that $\mathcal{H}M \in L^1(\mathbb{R})$ ([13, Chapter III, §7, Lemma 7.12]). We have $M \in L^2(\mathbb{R})$; hence $\mathcal{H}M \in L^2(\mathbb{R})$, more precisely, there holds

$$\|\mathcal{H}M\|_{L^2(\mathbb{R})} \le \|M\|_{L^2(\mathbb{R})},$$

since $C_2 = 1$ in (6.3). Now,

$$x\mathcal{H}M(x) = \frac{1}{\pi} \int_{\mathbb{R}} \frac{x}{x-t} M(t)\, dt$$

$$= \frac{1}{\pi} \int_{\mathbb{R}} M(t)\, dt + \frac{1}{\pi} \int_{\mathbb{R}} \frac{1}{x-t} t M(t)\, dt$$

$$= \mathcal{H}(tM(t))(x).$$

This yields

$$\|x\mathcal{H}M(x)\|_{L^2(\mathbb{R})} \le \|tM(t)\|_{L^2(\mathbb{R})};$$

hence, as above,

$$\|\mathcal{H}M\|_{L^1(\mathbb{R})} \le 2^{\frac{3}{2}} \|M\|_{L^2(\mathbb{R})}^{\frac{1}{2}} \|tM(t)\|_{L^2(\mathbb{R})}^{\frac{1}{2}}.$$

Like for atoms, there are other types of q-molecules which depend on a parameter q, with $q = 2$ in the considered case; however, they are of more importance for the H^p spaces with $p < 1$ rather than for H^1. Nevertheless, let us present the explicit value for the q-molecular norm (see, e.g., [13, Chapter III, §7]):

$$N_q(M) = \|M\|_{L^q(\mathbb{R})}^{\frac{1}{q}} \|tM(t)\|_{L^q(\mathbb{R})}^{\frac{1}{q'}},$$

where $N_2(M) := N(M)$. All the above reasoning goes along the same lines. Instead of repeating them, let us show how (5.27) can now be proved by the molecular approach. We assume $1 < q \le 2$, and in order to estimate, for the q-molecule M, the integral

$$\int_{\mathbb{R}} \left| \frac{\widehat{M}(x)}{x} \right| dx,$$

we split the integral into two: over $|x| \leq r$ and over $|x| > r$, with

$$r = \frac{\|M\|_{L^q(\mathbb{R})}}{\|tM(t)\|_{L^q(\mathbb{R})}}.$$

First, applying Hölder's inequality and then the Hausdorff-Young inequality with a rough constant, we obtain

$$\int_{|x|>r} \left| \frac{\widehat{M}(x)}{x} \right| dx \leq \left(\int_{|x|>r} |\widehat{M}(x)|^{q'} dx \right)^{\frac{1}{q'}} \left(\int_{|x|>r} |x|^{-q} dx \right)^{\frac{1}{q}}$$

$$\leq \|M\|_{L^q(\mathbb{R})} (2\pi)^{\frac{1}{q'}} \left(\frac{2}{q-1} \right)^{\frac{1}{q}} r^{-\frac{1}{q'}}$$

$$= (2\pi)^{\frac{1}{q'}} \left(\frac{2}{q-1} \right)^{\frac{1}{q}} N_q(M).$$

For the integral over $|x| \leq r$, we additionally split the integral in t into two: over $|t| \leq \frac{1}{|x|}$ and over $|t| > \frac{1}{|x|}$. In addition, for both, we estimate the Fourier integrals with $e^{-ixt} - 1$ rather than with e^{-ixt}, by the cancelation property. Using only Hölder's inequality, we get

$$\int_{|x| \leq r} \left| \int_{|t| > \frac{1}{|x|}} tM(t) \frac{e^{-ixt} - 1}{xt} dt \right| dx$$

$$\leq \int_{|x| \leq r} \int_{|t| > \frac{1}{|x|}} |tM(t)| \frac{dt}{|t|} \frac{dx}{|x|}$$

$$\leq \int_{|x| \leq r} \|tM(t)\|_{L^q(\mathbb{R})} \left(\int_{|t| > \frac{1}{|x|}} |t|^{-q'} dt \right)^{\frac{1}{q'}} \frac{dx}{|x|}$$

$$= 2q \left(\frac{2}{q'-1} \right)^{\frac{1}{q'}} N_q(M).$$

Finally, by the trivial estimate $\left| \frac{e^{-ixt}-1}{xt} \right| \leq 1$ and again by Hölder's inequality, we obtain

$$\int_{|x| \leq r} \left| \int_{|t| \leq \frac{1}{|x|}} tM(t) \frac{e^{-ixt} - 1}{xt} dt \right| dx$$

$$\leq \int_{|x| \leq r} \|tM(t)\|_{L^q(\mathbb{R})} \left(\int_{|t| \leq \frac{1}{|x|}} dt \right)^{\frac{1}{q'}} dx$$

$$= 2q 2^{\frac{1}{q'}} N_q(M).$$

Combining all these estimates, we get

$$\int_{\mathbb{R}} \left| \frac{\widehat{M}(x)}{x} \right| dx$$

$$\leq \left[(2\pi)^{\frac{1}{q'}} \left(\frac{2}{q-1} \right)^{\frac{1}{q}} + 2q \left(\frac{2}{q'-1} \right)^{\frac{1}{q'}} + 2q 2^{\frac{1}{q'}} \right] \|M\|_{L^q(\mathbb{R})}^{\frac{1}{q}} \|t M(t)\|_{L^q(\mathbb{R})}^{\frac{1}{q'}}.$$

With the Babenko–Beckner constant, the factor before $N_q(M)$ could be different but this does not seem essential and still it is not clear whether this factor will become sharp.

Again, with certain precautions, one can prove many results for the whole space by checking them only on molecules. To compare the two characterizations, recall that every atom is also a molecule. On the other hand, each molecule can be represented by means of a collection of atoms.

6.4 Subspaces

We will consider subspaces of the real Hardy space separately for the odd case and for the even one.

There is a scale of subspaces of $H_o^1(\mathbb{R}_+)$ proved to be convenient in various applications.

Definition 6.15 For $1 < q < \infty$, the space O_q is the space of functions g with finite norm

$$\|g\|_{O_q} = \int_0^\infty \left(\frac{1}{x} \int_x^{2x} |g(t)|^q dt \right)^{\frac{1}{q}} dx. \tag{6.16}$$

All these spaces and their sequence analogs first appeared in the paper by D. Borwein [78], but became—for sequences—widely known after the paper by G. A. Fomin [105]; see also [109]. On the other hand, these spaces are a partial case of the so-called Herz spaces (see first of all the initial paper by Herz [120] and a relevant paper of Flett [104]; see also [102] and [197]).

Further, for $q = \infty$ we define the corresponding space as follows.

Definition 6.17 The space O_∞ is the space of functions g with finite norm

$$\|g\|_{O_\infty} = \int_0^\infty \operatorname*{ess\,sup}_{x \leq t \leq 2x} |g(t)| \, dx. \tag{6.18}$$

We note that the functions in the integral in (6.18) is not necessarily monotone. However, its integrability or non-integrability is equivalent to that of the monotone

function ess sup $|g(t)|$. The role of the integrable monotone majorant for problems
$\underset{x \leq t}{}$
of almost everywhere convergence of singular integrals is known from the work
of D.K. Faddeev (see, e.g., [2, Chapter IV, §4]; also [51, Chapter I]); for spectral
synthesis problems it was used by Beurling [73], for more details see [70].

Of course, such spaces can be defined on the whole $\mathbb{R}$ by

$$\int_0^\infty \left(\frac{1}{x} \int_{x \leq |t| < 2x} |g(t)|^q dt \right)^{\frac{1}{q}} dx,$$

with obvious modification for $q = \infty$.

The basic interrelations between these spaces are given by the chain of embeddings

$$O_\infty \hookrightarrow O_{p_1} \hookrightarrow O_{p_2} \hookrightarrow H_o^1 \hookrightarrow L^1 \ (p_1 > p_2 > 1), \tag{6.19}$$

which can be found in [141], [24, Chapter 3], [151].

There is a convenient possibility to measure the norm in O_∞ in a different way.
By $L^* := L^*(\mathbb{R})$ we denote the space of functions f endowed with the norm

$$\|f\|_{L^*} = \int_0^\infty \underset{|t| \geq x}{\mathrm{ess\,sup}} |f(t)| \, dx < \infty. \tag{6.20}$$

It differs slightly from the norm in O_∞ (first of all, f is a function on the whole real
axis), but will be both convenient and used in the future.

Theorem 6.21 *We have* $f \in L^*$ *if and only if there exists a non-negative
function F locally absolutely continuous on* $(0, \infty)$ *and such that almost everywhere*
$|f(\pm x)| \leq F(x)$ *and* $F'(x) \leq \lambda(x)$ *for some positive function* λ *satisfying*

$$\int_0^\infty F(x) \, dx + \int_0^\infty x\lambda(x) \, dx < \infty.$$

To understand the properties of F and their relation to the property of monotonicity, let us formulate and prove the following important fact for such functions F.

Lemma 6.22 *For a function F satisfying the assumptions of Theorem 6.21, we have*

$$\lim_{x \to \infty} xF(x) = 0. \tag{6.23}$$

This is by no means true for an arbitrary integrable function F, but the additional
assumptions of the theorem weaker than the monotonicity guarantee (6.23). An
analogous statement for sequences was proved in [78, Theorem 1]; the proof for
functions goes along the same lines.

Proof We have, for $0 < x < y$,

$$\int_x^y t F'(t)\, dt \le \int_x^y t\lambda(t)\, dt.$$

This is equivalent to the inequality

$$y F(y) - x F(x) - \int_x^y F(t)\, dt \le \int_x^y t\lambda(t)\, dt,$$

and hence

$$x F(x) - y F(y) \ge -\int_x^y F(t)\, dt - \int_x^y t\lambda(t)\, dt. \qquad (6.24)$$

The right-hand side is negative and $o(1)$ as $x, y \to \infty$. Surely, $\lim\limits_{x \to \infty} x F(x)$ cannot be positive, otherwise $F(x)$ cannot be integrable. Hence there exists, for each positive ε, a sequence $\{x_n\}$ for which

$$x_n F(x_n) < \varepsilon.$$

Now suppose that $\overline{\lim\limits_{x \to \infty}}\, x F(x) > 0$. Then there is a sequence $\{y_k\}$ such that

$$y_k F(y_k) > 2\varepsilon.$$

For $\{x_n\}$ and $\{y_{k_n}\}$, we have $y_{k_n} > x_n$, and this contradicts (6.24), which yields (6.23). $\qquad\square$

Proof of Theorem 6.21 Given $f \in L^*$, a natural desire is to take

$$\operatorname*{ess\,sup}_{|t| \ge x} |f(t)|$$

as $F(x)$. This function is monotone decreasing but may be not locally absolutely continuous. The following very instructive result due to I. A. Shevchuk (private communication) solves this problem and shows that F can be chosen to be arbitrarily smooth.

Lemma 6.25 *For each monotone decreasing function g integrable on $(0, +\infty)$, arbitrary $\varepsilon > 0$, and any positive integer r, there is a monotone decreasing integrable function $G \in C^r(0, +\infty)$ such that*

$$G(x) \ge g(x) \qquad (6.26)$$

for each $x \in (0, +\infty)$ and

$$\int_0^\infty [G(x) - g(x)] \, dx < \varepsilon. \tag{6.27}$$

Proof of Lemma 6.25 Since g is monotone decreasing and integrable on $(0, +\infty)$, there exists a sequence of points $\{x_k\}_{k=-\infty}^{+\infty}$ such that $x_k < x_{k-1}$ for all k, $\lim_{k \to +\infty} x_k = 0$ while $\lim_{k \to -\infty} x_k = +\infty$, and

$$\sum_{k=-\infty}^{+\infty} g(x_k)(x_{k-1} - x_k) - \int_0^\infty g(x) \, dx < \frac{\varepsilon}{2}. \tag{6.28}$$

We denote by $t_k \in (x_k, x_{k-1})$ the points satisfying

$$\sum_{k=-\infty}^{+\infty} [g(x_{k+1}) - g(x_k)](t_k - x_k) < \frac{\varepsilon}{2}. \tag{6.29}$$

We observe that $t_k - x_k < x_k - x_{k+1}$. For each k, set

$$S_k(x) = \frac{\int_{x_k}^x (t_k - u)^r (u - x_k)^r \, du}{\int_{x_k}^{t_k} (t_k - u)^r (u - x_k)^r \, du}.$$

Finally, define G as

$$G(x) = \begin{cases} g(x_{k+1})(1 - S_k(x)) + g(x_k)S_k(x), & \text{if } \ x_k \le x \le t_k, \\ g(x_k), & \text{if } \ t_k < x \le x_{k-1}. \end{cases}$$

Obviously, (6.26) holds, and

$$G(x) - g(x) = I_1(x) + I_2(x),$$

where $I_1(x) = g(x_k) - g(x)$ on each $[x_k, x_{k-1}]$, and

$$I_2(x) = \begin{cases} [g(x_{k+1}) - g(x_k)](1 - S_k(x)), & \text{if } \ x_k \le x \le t_k, \\ 0, & \text{if } \ t_k < x \le x_{k-1}. \end{cases}$$

We obtain

$$\int_0^\infty I_2(x) \, dx = \sum_{k=-\infty}^{+\infty} (g(x_{k+1}) - g(x_k)) \int_{x_k}^{t_k} \frac{\int_x^{t_k} (t_k - u)^r (u - x_k)^r \, du}{\int_{x_k}^{t_k} (t_k - u)^r (u - x_k)^r \, du} \, dx,$$

and this value is bounded by the left-hand side of (6.29), while

$$\int_0^\infty I_1(x)\, dx$$

is exactly the left-hand side of (6.28). Hence we have (6.27), which completes the proof. □

Back to the proof of the theorem, considering

$$g(x) = \operatorname{ess\,sup}_{|t| \geq x} |f(t)|,$$

we may take the corresponding G for F. By this, we may take

$$\lambda(x) = -F'(x).$$

Then $\int_0^\infty F(x)\, dx < \infty$, and

$$\int_0^\infty F(x)\, dx = x F(x) \Big|_0^\infty - \int_0^\infty x F'(x)\, dx$$

$$= \int_0^\infty x \lambda(x)\, dx < \infty.$$

Let us prove the converse statement. Given a function F satisfying the assumptions of the theorem, let us show that $F \in L^*$. Since

$$\operatorname{ess\,sup}_{|t| \geq x} |f(t)| \leq \operatorname{ess\,sup}_{|t| \geq x} F(t) = \Phi(x),$$

this will definitely prove the theorem. We have

$$\int_0^\infty \Phi(x)\, dx = \sum_{n=1}^\infty F(x_n)(x_n - a_n) + \sum_{n=1}^\infty \int_{x_{n-1}}^{a_n} F(x)\, dx,$$

where $x_0 = 0$, $a_n > x_{n-1}$ for $n > 1$, and $a_1 \geq x_0$. Here (x_{n-1}, a_n) are the intervals where $F(x)$ is monotone decreasing. Observe that $F(x_n) = F(a_n)$. We obtain

$$F(x_n)(x_n - a_n) = \int_{a_n}^{x_n} F(x)\, dx + \int_{a_n}^{x_n} x F'(x)\, dx,$$

and therefore

$$
\int_0^\infty \Phi(x)\,dx = \int_0^\infty F(x)\,dx + \sum_{n=1}^\infty \int_{a_n}^{x_n} x\,F'(x)\,dx
$$

$$
\le \int_0^\infty F(x)\,dx + \int_0^\infty x\lambda(x)\,dx < \infty. \tag{6.30}
$$

Hence $\Phi \in L^*$, and the theorem is proved. $\square$

For more details, relations with integrability of trigonometric series and the history of this specific result, see [143].

Now, we present a similar scale of nested spaces being subspaces of $H_e^1(\mathbb{R}_+)$. The construction comes from (5.25) and from the structure of O_q.

Definition 6.31 We define, for $1 < q \le \infty$,

$$
E_q := E_q(\mathbb{R}_+) := \left\{ g \in O_q : \int_0^\infty \frac{1}{x} \left| \int_0^x g(t)\,dt \right| dx < \infty \right\}.
$$

The sum of the last integral and the norm in O_q gives the norm in E_q:

$$
\|g\|_{E_q} = \int_0^\infty \left(\frac{1}{x} \int_x^{2x} |g(t)|^q\,dt \right)^{\frac{1}{q}} dx + \int_0^\infty \frac{1}{x} \left| \int_0^x g(t)\,dt \right| dx < \infty.
$$

In addition to the mentioned above, we emphasize that both scales of spaces proved to be important in the estimates of the Fourier transforms of functions with bounded variation; see [34].

6.5 A Paley–Wiener Theorem

The argument in Example 5.40 and some other examples of the behavior of a function near infinity might be completely irrelevant in the case of odd functions. Indeed, by a Paley–Wiener theorem, if an odd integrable function is monotone decreasing (or general monotone in some sense, see [165] and [164]), no matter how slow, then its Hilbert transform is integrable [177]. This is not the case for even functions. The difference apparently comes from the fact that an odd function automatically has mean zero. This allows such an $H^1(\mathbb{R})$ function (more precisely, an $H_o^1(\mathbb{R}_+)$ function) to be of one sign on the half-axis.

The following matter is related to both this section and the previous one. It gives us a nice opportunity to again illustrate how the atomic characterization works and how effective it may be.

Let us first outline this problem in a more general context. It is well known that for the Hilbert transform $\mathcal{H}g(x)$ and the weight

$$w(x) = |x|^{\alpha}, \quad \text{with} \quad -1 < \alpha < p - 1,$$

there holds

$$\|\mathcal{H}g\|_{L_w^p} \lesssim \|g\|_{L_w^p}, \quad 1 < p < \infty.$$

Here $L_w^p := L_w^p(\mathbb{R})$ means the *weighted Lebesgue space* endowed with the norm (cf. (7.1))

$$\|g\|_{L_w^p} = \|g\|_{L_w^p(\mathbb{R})} = \left(\int_{\mathbb{R}} |g(t)|^p w(t) \, dt \right)^{\frac{1}{p}}, \tag{6.32}$$

where the weight w is a non-negative locally integrable function. In [117], Hardy and Littlewood showed that for even functions g, this inequality also holds for

$$-p - 1 < \alpha < p - 1.$$

Later, Flett [103] proved the same results for odd functions provided $-1 < \alpha < 2p - 1$. For $p = 1$, it is known that only weak type inequality (5.13) holds for the Hilbert transform in general (see the previous chapter). On the other hand, Paley–Wiener's theorem asserts that for an odd and monotone decreasing function $g \in L^1$ on $\mathbb{R}_+$, one has $\mathcal{H}g \in L^1$. In [165], this theorem was extended to general monotone functions. Further, in [164] the weighted analogues of the Paley–Wiener theorem for odd and even (general monotone) functions were proved. In other words, it was an extension of Hardy–Littlewood's [117], Flett's [103] and Andersen's [62] results to the case $p = 1$ under the assumption of (generalized) monotonicity for an even/odd function.

Besides the initial proof in [177] (for series) and additional study in [202], a different proof of the initial Paley–Wiener theorem can be found in [49, Chapter IV, 6.2]. Let us give details. First of all, integrability of the Hilbert transform of an integrable function means that this function belongs to the real Hardy space $H^1(\mathbb{R})$.

Let g_0 be a non-negative monotone decreasing function on $(0, \infty)$ such that

$$\int_0^\infty g_0(t) \, dt < \infty,$$

and let $g(t) = g_0(t)$ on $(0, \infty)$, and $g(-t) = -g(t)$. Therefore, the Paley–Wiener theorem then states that $g \in H^1(\mathbb{R})$. The proof in [49, Chapter IV, 6.2] goes along the following lines. For $-\infty < k < \infty$, let

$$a_k(t) = \frac{1}{2^{k+2} g_0(2^k)} g_0(|t|) \operatorname{sign} t$$

if $2^k \leq |t| < 2^{k+1}$ and zero otherwise. Obviously, each a_k is an atom (in fact, a $(1, \infty, 0)$ atom). To see that the absolute value of the function is less than the reciprocal of the length of the support interval $[-2^{k+1}, 2^{k+1}]$, just the monotonicity is used. Taking $\lambda_k = 2^{k+2} g_0(2^k)$ and observing that

$$\sum_{k=-\infty}^{\infty} \lambda_k \leq 8 \int_0^{\infty} g_0(t) \, dt,$$

we see that the series $\sum_{k=-\infty}^{\infty} \lambda_k a_k(t)$ converges to $g(t)$ except at the origin. Since thus we have an atomic decomposition of g, it belongs to $H^1(\mathbb{R})$.

We can immediately extend both this result and its proof by taking g to be *weak monotone*. To define the latter notion, we will assume a function to lie on $(0, \infty)$, to be locally of bounded variation, and vanishing at infinity.

Definition 6.33 We say that a non-negative function f defined on $(0, \infty)$, is *weak monotone*, written WM, if

$$f(t) \leq Cf(x) \quad \text{for any} \quad t \in [x, 2x]. \tag{6.34}$$

Using in the above proof $g_0(t) \leq Cg_0(2^k)$ provided $g_0 \in WM$ instead of $g_0(t) \leq g_0(2^k)$ for monotone g_0, we immediately arrive at the following more general result than Theorem 6.1 in [165], since the notion of weak monotonicity introduced and widely used in [166] is less restrictive than certain notions of general monotonicity (see [165]).

Theorem 6.35 *Let g_0 be integrable on $(0, \infty)$ and $g_0 \in WM$. Then $g \in H^1(\mathbb{R})$, or, equivalently, its Hilbert transform is integrable.*

This shows that for the integrability of the Hilbert transform smoothness conditions are frequently not of crucial importance; certain regularity of the functions works instead.

6.6 Discrete Hardy Spaces

As mentioned (see Sect. 5.5 in Chap. 5), the discrete Hardy spaces will be denoted by h, h_e and h_o and will mean, for an ℓ^1 sequence a, that the corresponding discrete Hilbert transform $\hbar a$, $\hbar_e a$ or $\hbar_o a$ is also in ℓ^1. There are works where the analogs of the above results are established; see, e.g., [80] or [152] and references therein. They are real counterparts of the continuous variants, therefore we shall mainly follow [61], where more specific results are obtained. Of course, there are continuous predecessors for them as well, see, e.g., [170] and references therein.

Here, it will be convenient for us to denote by $|E|$ the number of lattice points contained in $E \subset \mathbb{Z}$. The distribution function of the discrete Hilbert transform (see interesting results for the distribution function of usual Hilbert transform in [170] and references therein)

$$\mathcal{D}\hbar a(s) = |\{n \in \mathbb{Z} : |\hbar a(n)| > s\}| = \sum_{n \in \mathbb{Z} : |\hbar a(n)| > s} 1 \tag{6.36}$$

is known to satisfy the weak type $(1, 1)$ inequality (see [124])

$$|\mathcal{D}\hbar a(s)| \leq \frac{C}{s} \|a\|_{\ell^1}. \tag{6.37}$$

A useful limit relation is proved in [61] for this distribution function as $s \to 0+$.

Theorem 6.38 *Let* $a \in \ell^1$. *Then the equation*

$$\lim_{s \to 0+} s\mathcal{D}\hbar a(s) = 2 \left| \sum_{k \in \mathbb{Z}} a_k \right| \tag{6.39}$$

holds.

Proof The proof starts with an auxiliary lemma.

Lemma 6.40 *Let* $a \in \ell^1$ *and*

$$\sum_{k \in \mathbb{Z}} a_k = 0. \tag{6.41}$$

Then the estimation

$$\mathcal{D}\hbar a(s) = o\left(\frac{1}{s}\right) \tag{6.42}$$

holds as $s \to 0+$.

Proof of the Lemma For a moment, we assume that the sequence in question is of bounded support, that is, has non-trivial values on an interval $[-K, K]$, with $a_k = 0$ if $|k| > K$. We then have for $|n| > K$,

$$\hbar a(n) = \sum_{|k| \leq K} \frac{a_k}{n - k} - \frac{1}{n - \frac{1}{2}} \sum_{|k| \leq K} a_k$$

$$= \sum_{|k| \leq K} \frac{k - \frac{1}{2}}{(n - k)(n - \frac{1}{2})} a_k,$$

which implies for large n, say, $|n| > 2K$,

$$|\hbar a(n)| \leq \frac{4}{n^2} \sum_{|k| \leq K} \left| k - \frac{1}{2} \right| |a_k|.$$

This yields (6.42) in the considered case. Indeed, all $n > \frac{1}{s}$ do not take part in forming $D\hbar a(s)$, while the number of the rest of n is finite (depending on K). By this, $sD\hbar a(s) \lesssim s$, as desired.

Proceeding to the general case, we observe that (6.41) implies that for any $\varepsilon > 0$, there holds $a = \alpha + \beta$, where α and β are two ℓ_1 sequences, with α finite (concentrated on $[-K, K]$ for some K) and satisfying (6.41), and $\|\beta\|_{\ell_1} < \varepsilon$. It follows from Lemma 6.40 that there exists S such that for all $0 < s < S$,

$$sD\hbar\alpha\left(\frac{s}{2}\right) < \varepsilon.$$

Further, it follows from (6.37) that for the same s, we have

$$sD\hbar\beta\left(\frac{s}{2}\right) \lesssim \|\beta\|_{\ell_1} \lesssim \varepsilon.$$

From these two inequalities and

$$\{n \in \mathbb{Z} : |\hbar a(n)| > s\} \subset \left\{n \in \mathbb{Z} : |\hbar\alpha(n)| > \frac{s}{2}\right\} \bigcup \left\{n \in \mathbb{Z} : |\hbar\beta(n)| > \frac{s}{2}\right\},$$

relation (6.42) follows.   □

Now, the theorem follows from this lemma provided (6.41) holds. Let us assume that

$$\sum_{k \in \mathbb{Z}} a_k = A \neq 0.$$

We again represent $a = \alpha + \beta$ but now the sequence α coincides with a, that is, $\alpha_k = a_k$ everywhere except $k = 0$, where $\alpha_0 = a_0 - A$, while $\beta_k = 0$ everywhere, with $\beta_0 = A$. Since α satisfies (6.41), Lemma 6.40 yields

$$D\hbar\alpha(s) = o\left(\frac{1}{s}\right), \qquad s \to 0+.$$

By definition, $\hbar\beta(n) = \frac{A}{n}$ if $n \neq 0$, and $\hbar\beta(0) = 0$. Therefore,

$$D\hbar\beta(s) \sim \frac{2|A|}{s}, \qquad s \to 0+.$$

These two relations and

$$\{n \in \mathbb{Z} : |\hbar\beta(n)| > (1 + \varepsilon)s\} \setminus \{n \in \mathbb{Z} : |\hbar\alpha(n)| > \varepsilon s\}$$

$$\subset \{n \in \mathbb{Z} : |\hbar a(n)| > s\}$$

$$\subset \{n \in \mathbb{Z} : |\hbar\alpha(n)| > \varepsilon s\} \bigcup \{n \in \mathbb{Z} : |\hbar\beta(n)| > (1 - \varepsilon)s\},$$

valid for any $0 < \varepsilon < 1$, yield

$$\frac{2|A|}{1 + \varepsilon} \leq \liminf_{s \to 0+} s\mathcal{D}\hbar a(s) \leq \limsup_{s \to 0+} s\mathcal{D}\hbar a(s) \leq \frac{2|A|}{1 - \varepsilon}.$$

This obviously implies (6.39), as required.   □

This theorem is not only of interest in its own right but allows one to prove the discrete analog of (5.16).

Theorem 6.43 *If $a \in h$, then (6.41) holds true.*

Proof Similarly to above, the distribution function of the sequence a is

$$\mathcal{D}a(s) = |\{n \in \mathbb{Z} : |a_n| > s\}|.$$

Let us prove that if a is summable, then

$$\mathcal{D}a(s) = o\left(\frac{1}{s}\right), \qquad s \to 0+. \tag{6.44}$$

We have

$$\sum_{k \in \mathbb{Z}} |a_k| = \sum_{k \in \mathbb{Z}: |a_k| > 1} |a_k| + \sum_{m=0}^{\infty} \sum_{\substack{k \in \mathbb{Z}: \\ |a_k| \in (2^{-m-1}, 2^{-m}]}} |a_k|$$

$$\geq |\{k \in \mathbb{Z} : |a_k| > 1\}| + \sum_{m=0}^{\infty} 2^{-m-1} |\{k \in \mathbb{Z} : |a_k| \in (2^{-m-1}, 2^{-m}]\}|$$

$$= \mathcal{D}a(1) + \sum_{m=0}^{\infty} 2^{-m-1} [\mathcal{D}a(2^{-m-1}) - \mathcal{D}a(2^{-m})]$$

$$= \sum_{m=0}^{\infty} 2^{-m-1} \mathcal{D}a(2^{-m}),$$

which implies

$$\lim_{m \to \infty} 2^{-m} \mathcal{D}a(2^{-m}) = 0.$$

Since $\mathcal{D}a(s)$ is decreasing as $s \to 0+$, relation (6.44) follows. For $\hbar a \in \ell^1$, changing a to $\hbar a$ in (6.44) turns it to

$$\mathcal{D}\hbar a(s) = o\left(\frac{1}{s}\right), \qquad s \to 0+ .$$

Along with (6.39), we obtain (6.41).   □

There also holds an analog of the Zygmund-Stein $L \log L$ condition.

Theorem 6.45 *If the sequence $a \in \ell^1$ satisfies (6.41) and*

$$\sum_{k \in \mathbb{Z}} |a_k| \ln(e + |k|) < \infty, \tag{6.46}$$

then $\hbar a \in \ell^1$ and

$$\|a\|_h = \|a\|_{\ell^1} + \|\hbar a\|_{\ell^1} \lesssim \sum_{k \in \mathbb{Z}} |a_k| \ln(e + |k|).$$

Proof By definition, we have

$$|\hbar a(0)| = \left| \sum_{k \neq 0} \frac{a_k}{k} \right| \leq \|a\|_{\ell^1}.$$

For $n \neq 0$, it follows from (6.41) that

$$|\hbar a(n)| = \left| \sum_{k \neq n} \frac{a_k}{n - k} \right| = \left| \sum_{k \neq n} \frac{a_k}{n - k} - \sum_{k \neq n} \frac{a_k}{n} - \frac{a_n}{n} \right|$$

$$\leq \left| \frac{a_n}{n} \right| + \sum_{k \neq n} \frac{|k|\, |a_k|}{|n|(|n - k|)}.$$

These two inequalities imply that

$$\|\hbar a\|_{\ell^1} = \sum_{n \in \mathbb{Z}} |\hbar a(n)| \leq \|a\|_{\ell^1} + \sum_{n \neq 0} \sum_{k \neq n} \frac{|k|\, |a_k|}{|n|\, |n - k|}$$

$$\leq \|a\|_{\ell^1} + 2 \sum_{n=1}^{\infty} \left(\sum_{k > n} + \sum_{1 \leq k < n} \right) \frac{k\, |a_k|}{n\, |n - k|}$$

$$+ 2 \sum_{n=1}^{\infty} \left(\sum_{k > n} + \sum_{1 \leq k < n} \right) \frac{k\, |a_k|}{n\, (n + k)}$$

$$\lesssim \sum_{k \in \mathbb{Z}} |a_k| \ln(e + |k|),$$

which completes the proof.   □

Naturally, under assumptions of Theorem 6.43 there also holds (see Theorem 4 in [61])

$$\sum_{k\in\mathbb{Z}} \hbar a(k) = 0.$$

We mention that a discrete version of the above Paley–Wiener theorem also exists, see [152].

6.7 Back to Trigonometric Series

In this section, we return to the topic of trigonometric series, already considered in Sect. 3.6 of Chap. 3, not because of different results (though some of them, very recent, will be given) but more because of different methods. This became possible after the introduction of new notions and terminology. To start with, recall the so-called Boas-Telyakovskii condition (3.40). It can be replaced by the assumption that the sequence of differences of the coefficients belongs to the discrete Hardy space h_o. This turns out to be not only intrinsic language but a way to indicate what kind of tools should be involved in this study.

A traditional way to prove such results were the so-called Szidon type inequalities mentioned in Sect. 3.6 of Chap. 3. In [141], a new approach to these problems was introduced.

First, let us consider a locally absolutely continuous function f on $(0, \infty)$ such that

$$\lim_{x\to\infty} f(x) = 0$$

and $f' \in X$, where X is a subspace of the space of integrable functions. For example, this is the case for functions of bounded variation, locally absolutely continuous on $(0, \infty)$ and vanishing at infinity; we can denote (cf. Theorem 2.21 in Chap. 2)

$$\|f\|_{BV} = \int_0^\infty |f'(x)| \, dx < \infty.$$

Various spaces X are considered in detail in [34]; roughly speaking, most of them are non-periodic analogs of the sequence spaces considered earlier in the sense that the sequences are replaced by functions and the sums by integrals. One of these spaces is the real Hardy space $H^1(\mathbb{R})$. We describe how it comes into play in a natural way. For the given class of functions, we integrate by parts in the formula for the Fourier transform (4.1) understood in the improper sense. What we arrive at instead is

$$\frac{\widehat{f'}(x)}{ix}. \tag{6.47}$$

Integrating $|\widehat{f}|$ over $\mathbb{R}$ is the same as integrating over $\mathbb{R}$ the absolute value of (6.47). But, by the Fourier-Hardy inequality (5.27), the latter is dominated by the norm of f' in $H^1(\mathbb{R})$. This brief and transparent argument already gives the following strong condition for the integrability of the Fourier transform.

Theorem 6.48 *If a function f is locally absolutely continuous on $\mathbb{R} \setminus \{0\}$,*

$$\lim_{|t| \to \infty} f(t) = 0,$$

and $f' \in H^1(\mathbb{R})$, then the Fourier transform $\widehat{f}$ is Lebesgue integrable on $\mathbb{R}$.

More sophisticated results, including asymptotic ones, are collected, as mentioned, in [34]; see also references therein. It is of considerable interest that in Theorem 6.48 there is no need to assume that f is of bounded variation. Indeed, what we need the boundedness of variation for is the differentiability of the derivative. But here we get this from the assumption $f' \in H^1(\mathbb{R})$.

Let us go on to the Fourier transform approach. In order to present an example of a theorem on the behavior of the Fourier transform, we consider $\widehat{f_c}$ and $\widehat{f_s}$ as in (4.5) and (4.6), respectively.

Theorem 6.49 *Let f be as above. Then for any $y > 0$ we have*

$$\widehat{f_c}(y) = \theta \gamma_1(y),$$

and

$$\widehat{f_s}(y) = y^{-1} f\left(\frac{\pi}{2y}\right) + \theta \gamma_2(y),$$

where $|\theta| \leq C$ and

$$\int_0^\infty |\gamma_j(y)| \, dy \leq \|f'\|_X, \qquad j = 1, 2.$$

With this or similar theorems in hand (again, they can be found first of all in [34]), we are now able to show a way how to strengthen the known integrability results. Given cosine and sine series (3.35) and (3.36), with the null sequence of coefficients $\{a_k\}$ and $\{b_k\}$ such that their difference sequences $\{\Delta a_k\}$ and $\{\Delta b_k\}$,

$$\Delta a_k = a_k - a_{k+1} \qquad \text{and} \qquad \Delta b_k = b_k - b_{k+1},$$

are in a sequence space χ (acting in the same sense for sequences as X above for functions). For this, a "bridge" for passing from Fourier integrals to trigonometric series (and vice versa) is needed. In the first one, in [68], the function f was assumed to be of bounded variation (and of compact support which is an unnecessary restriction). Further results are due to Trigub, see [196] and [55, Theorem 4.1.2]. For the function f of bounded total variation and vanishing at infinity, it reads as follows:

$$\sup_{|x|\leq\pi} \left| \int_{\mathbb{R}} f(t)e^{ixt}\,dt - \sum_{k\in\mathbb{Z}} f(k)e^{ixk} \right| \lesssim \|f\|_{BV}. \tag{6.50}$$

With this in hand, we set for $x \in [k, k+1]$,

$$a(x) = a_k + (k-x)\Delta a_k, \qquad a_0 = 0,$$

$$b(x) = b_k + (k-x)\Delta b_k.$$

So, we construct a corresponding function by means of linear interpolation of the sequence of coefficients. The same can be expressed in a somewhat different manner, even in a more general form. We follow Lemma D in [114]. Let

$$\Delta(x) = \begin{cases} 1 - |x|, & |x| \leq 1, \\ 0, & |x| > 1. \end{cases}$$

Lemma 6.51 *Let $\{c_k\}$ be a sequence of complex numbers and let*

$$\ell_c(x) = \sum_{k=-\infty}^{\infty} c_k \Delta(x-k), \qquad x \in \mathbb{R}. \tag{6.52}$$

Then $\ell_c(k) = c_k, k \in \mathbb{Z}$, and ℓ_c is linear on each interval $[k, k+1]$.

In other words, this lemma says that the function whose graph consists of the line segments joining (k, c_k) to $(k+1, c_{k+1})$ for $k = 0, \pm1, \pm2, \ldots$ is given by the series (6.52).

Starting with assumptions on the coefficients as belonging to χ, we arrive at

Theorem 6.53 *For each y, where $0 < y \leq \pi$, we have*

$$\sum_{k=1}^{\infty} a_k \cos ky = \theta\gamma(y)$$

and

$$\sum_{k=1}^{\infty} b_k \sin ky = \frac{1}{y} b\left(\frac{\pi}{2y}\right) + \theta\gamma(y),$$

where $|\theta| \le C$ and

$$\int_0^\pi |\gamma(y)|\, dy \le \|\{\Delta a_k\}\|_\chi$$

for the first relation and similarly for b.

Recently [153], a counterpart of Theorem 6.49 has been obtained, in the sense that instead of h_o as χ the space h_e was taken. This resulted in an asymptotic formula for $\widehat{f_c}$ rather than for $\widehat{f_s}$. This result was completely new, that is, never obtained before directly for sequence analogs. It came as a consequence of similarly new results for trigonometric series. The corresponding theorem (for the following somewhat more precise form, see [34, Chapter 8, Theorem 8.8]) reads as follows.

Theorem 6.54 *If the coefficients $\{a_k\}$ in (3.35) and $\{b_k\}$ in (3.36) tend to 0 as $k \to \infty$, and the sequences $\{\Delta a_k\}$ and $\{\Delta b_k\}$ are in h_e, then (3.36) represents an integrable function on $[0, \pi]$, and, for $N \le \frac{\pi}{2x} < N + 1$,*

$$\sum_{k=1}^\infty a_k \cos kx = \frac{A}{x} a\left(\frac{\pi}{2x}\right) + \frac{\pi}{2x} \sum_{k=1}^{N-1} \Delta a_k \ln \frac{e\pi}{2x(k+1)(1+\frac{1}{k})^k} + G(x),$$

where

$$A = \frac{2}{\pi}\left(\int_0^{\frac{\pi}{2}} \frac{1 - \cos t}{t}\, dt - \int_{\frac{\pi}{2}}^\infty \frac{\cos t}{t}\, dt\right)$$

and

$$\int_0^\pi |G(x)|\, dx \lesssim \|\{\Delta a_k\}\|_{h_e}.$$

The above approach does not cover all the options because of the reduction to functions and sequences of bounded variation or those related to them. Recently, such a restriction has been removed (or, more precisely, the range of the Fourier transform approach has been extended) by finding in [168] an optimal necessary and sufficient condition for a trigonometric series to be a Fourier series (in fact, the same can be found in an earlier work [114]; the reason that this and similar results in [114] have gone unnoticed by the researchers of trigonometric series for a long time is probably due to the absence of applications in that paper). In the comprehensive case, there is less reason to distinguish between the cosine and sine series, and the general complex form (3.2) is more appropriate. Similarly to above, the answer is given in terms of a single, quite simple function ℓ_c, piecewise linear, continuous and such that $\ell_c(k) = c_k, k \in \mathbb{Z}$, that is, exactly like in Lemma 6.51.

Criterion *For series (3.2) to be a Fourier series, it is necessary and sufficient that* $\ell_c \in W_0(\mathbb{R})$.

It is hardly believable that a necessary and sufficient condition closer to the initial form of the series (that is, in terms of $\{c_k\}$ only) may exist. This criterion results in two ways for obtaining new and known conditions for the trigonometric series to be a Fourier series. One of them is to search for specific conditions for the function ℓ_c to be in W_0. Such an approach leaves a chance for discoveries but a general feeling is that in this case any calculations will be equivalent to those with the sequence $\{c_k\}$, just because of the structure of ℓ_c. More promising seems to be the application of general conditions for belonging to W_0 to the concrete function ℓ_c. In any case, in [168] completely new results not related to bv immediately appeared as consequences of earlier obtained results for functions to belong to W_0. For example, one of them comes as a consequence of Theorem 8.30 to be discussed in the last chapter.

Theorem 6.55 *If there is* $p \in (0, 2]$ *such that* $\{c_k\} \in l_p$, *then (3.2) is the Fourier series of an* L_2 *function. If* $\{c_k\} \in l_p$ *only for some* $p \in (2, +\infty)$, *then assuming* $\{c_k - c_{k+1}\} \in l_q$, *with* $q \in \left(0, 1 + \frac{1}{p-1}\right)$, *we get that (3.2) is a Fourier series.*

Chapter 7
Hardy Inequalities

In this chapter, we return to the issue slightly touched on in Chap. 5. A version of the celebrated Hardy inequality has been given by (5.28). Hardy's inequalities are an important part of analysis and frequently used tools. They exist in various forms, they are the subject of numerous books (see, e.g., [32, 33, 42]). We mention that (5.28) is more or less a historical fact, since there are many much more advanced versions. More precisely, let $L_v^p := L_{v(\cdot)}^p(Y)$, with

$$\|f\|_{p,v} = \left(\int_Y |f(y)|^p v(y)^p \, dy \right)^{\frac{1}{p}}, \tag{7.1}$$

be the weighted Lebesgue space, defined in a slightly different way than in (6.32). In our considerations $Y = \mathbb{R}_+ = [0, \infty)$. The weight v (and u) is a non-negative locally integrable function.

Some of the most general Hardy's inequalities with general weights are as follows (see [81]):

$$\|P_x f\|_{q,u} \lesssim \|f\|_{p,v} \tag{7.2}$$

and

$$\|Q_x f\|_{q,u} \lesssim \|f\|_{p,v}, \tag{7.3}$$

with $f \geq 0$, $1 \leq p \leq q < \infty$, where the constants are independent of functions f and

$$P_x f = \int_0^x f(y) \, dy \quad \text{and} \quad Q_x f = \int_x^\infty f(y) \, dy \tag{7.4}$$

© The Author(s), under exclusive license to Springer Nature Switzerland AG 2021
E. Liflyand, *Harmonic Analysis on the Real Line*, Pathways in Mathematics,
https://doi.org/10.1007/978-3-030-81892-0_7

being the Hardy and Bellman operators, respectively. Inequality (7.2) holds if and only if, for each $r > 0$,

$$\left(Q_r u^q\right)^{\frac{1}{q}} \left(P_r v^{-p'}\right)^{\frac{1}{p'}} \lesssim 1, \tag{7.5}$$

where here and in similar conditions the constant in $\lesssim$ on the right does not depend on r. For $p = 1$ and $q < \infty$, see also [42, (5.12)]. Similarly, (7.3) holds if and only if, for each $r > 0$,

$$\left(P_r u^q\right)^{\frac{1}{q}} \left(Q_r v^{-p'}\right)^{\frac{1}{p'}} \lesssim 1. \tag{7.6}$$

To illustrate a variety of possibilities for inequalities to be considered as Hardy's inequalities, let us mention the following Steklov–Hardy type inequality (see [33, Cor.3.14]):

For $F \geq 0$, weights U and V, and $1 < q \leq Q < \infty$, we have

$$\left(\int_{\mathbb{R}} \left[\int_{t-h}^{t+h} F(s)\, ds\right]^Q U(t)\, dt\right)^{\frac{1}{Q}} \leq \gamma(h) \left(\int_{\mathbb{R}} F^q(t) V(t)\, dt\right)^{\frac{1}{q}} \tag{7.7}$$

if and only if there holds

$$A := \sup_{t \leq u \leq 2h+t} \left(\int_t^u U(s)\, ds\right)^{\frac{1}{Q}} \left(\int_{u-h}^{t+h} V^{1-q'}(s)\, ds\right)^{\frac{1}{q'}} < \infty. \tag{7.8}$$

Moreover, $\gamma(h) \leq CA$. We will use it in the next chapter.

Let us also note that in parallel to the functional setting, sequence versions have been intensively studied. It can be said that for most of the known functional inequalities, there are counterparts for sequences and vice versa. It is not our goal to substitute for the mentioned sources or to give a complete picture of the topic. We restrict ourselves to two issues. First, we begin with the sequence case. However, rather than prove one of the most general variants, we will present a very recent, very brief and impressive proof of a not that advanced but classical sequence version in [139]. Further, in the functional setting, we will prove a yet more extended version of the very general classical weighted version due to Bradley [155]. These generalizations will introduce us into the world of the so-called Hausdorff operators, a topic not really touched on in the preceding study.

All together, this chapter is a transition chapter between the previous "theoretical" matter and applications. Here they are united.

7.1 Discrete Hardy Inequality

As mentioned, we do not try to present the most general sequence versions of Hardy's inequalities as in the following section for functions. The inequality we shall deal with is as follows. Recall that in (4.28) the norm of the sequence $\{d_j\}$ in ℓ^r is defined. We rewrite it starting from $j = 0$ rather than $j = 1$ in (4.28):

$$\|\{d_j\}\|_{\ell^r} = \left(\sum_{j=0}^{\infty} |d_j|^r\right)^{\frac{1}{r}}.$$

With any sequence $\{d_j\}$, we associate the sequence

$$A_n = \frac{1}{n+1}\sum_{j=0}^{n} d_j.$$

The celebrated discrete Hardy inequality reads as follows.

Let $p > 1$. For every $\{d_j\} \in \ell^p$, we have

$$\left(\sum_{n=0}^{\infty}\left|\frac{1}{n+1}\sum_{j=0}^{n} d_j\right|^p\right)^{\frac{1}{p}} \le p'\left(\sum_{j=0}^{\infty}|d_j|^p\right)^{\frac{1}{p}}. \tag{7.9}$$

Here the constant p', which is sharp, is also of importance. This is not the case in some applications but very often just the search for the sharp constant is the main difficulty. There are many classical proofs of both the inequality itself and the sharpness of the constant. Instead we present a very recent and probably "the most direct" proof of (7.9) due to Lefèvre [139].

Proof First, it is well known that it suffices to prove the inequality for $\{d_j\}$ non-increasing. Indeed, the ℓ^p norm of $\{d_j\}$ is the same for the non-increasing rearrangement (see (2.4) and further) and $\|\{A_n\}\|_{\ell^p} \le \|\{A_n^*\}\|_{\ell^p}$. For every integer $n \ge 0$, we can write

$$A_n = \sum_{j=0}^{n}\int_{\frac{j}{n+1}}^{\frac{j+1}{n+1}} d_j \, ds = \int_0^1 d_{[(n+1)s]} \, ds,$$

where $[\alpha]$ is, as usual, the integer part of α. By the Minkowski inequality, we obtain

$$\left(\sum_{n=0}^{\infty}\left|\frac{1}{n+1}\sum_{j=0}^{n} d_j\right|^p\right)^{\frac{1}{p}} \le \int_0^1\left(\sum_{n=0}^{\infty}|d_{[(n+1)s]}|^p\right)^{\frac{1}{p}} ds$$

$$\le \int_0^1\left(\sum_{n=1}^{\infty}|d_{[ns]}|^p\right)^{\frac{1}{p}} ds.$$

For every $s \in (0, 1)$ and for $m \in \mathbb{N} \cup \{0\}$, we introduce

$$I_m(s) = \{n \geq 1 : [ns] = m\} = \left[\frac{m}{s}, \frac{m+1}{s}\right) \bigcap \mathbb{N}.$$

Clearly, $\{I_m\}$ is a partition of $\mathbb{N}$, so we have

$$\sum_{n=1}^{\infty} |d_{[ns]}|^p = \sum_{m=0}^{\infty} \sum_{n \in I_m(s)} |d_{[ns]}|^p = \sum_{m=0}^{\infty} |d_m|^p |I_m(s)|, \qquad (7.10)$$

where here again $|E|$ means the number of points in the discrete set E. For every irrational $s \in [0, 1]$, and for every $m \geq 0$,

$$|I_m(s)| = \left[\frac{m+1}{s}\right] - \left[\frac{m}{s}\right]. \qquad (7.11)$$

We now need the following auxiliary relation.

Lemma 7.12 *Let $\{\alpha_m\} \in \ell^1$ be a non-increasing summable sequence of non-negative real numbers and $\lambda > 0$. Then*

$$\sum_{m=0}^{\infty} \alpha_m([(m+1)\lambda] - [m\lambda]) \leq \lambda \sum_{m=0}^{\infty} \alpha_m.$$

Proof Applying Abel's transformation (see (3.37) in Chap. 3), we obtain

$$\sum_{m=0}^{\infty} \alpha_m([(m+1)\lambda] - [m\lambda]) = \sum_{m=0}^{\infty} [(m+1)\lambda](\alpha_m - \alpha_{m+1})$$

$$\leq \lambda \sum_{m=0}^{\infty} (m+1)(\alpha_m - \alpha_{m+1}),$$

since $\{\alpha_m\}$ is non-increasing and $[k\lambda] \leq k\lambda$ for every $k \geq 0$. Applying again Abel's transformation, in the backward direction, we complete the proof. □

We now complete the proof of the inequality as follows. Combining (7.10), (7.11) and Lemma 7.12 yields

$$\left(\sum_{n=1}^{\infty} |d_{[ns]}|^p\right)^{\frac{1}{p}} \leq s^{-\frac{1}{p}} \|\{d_j\}\|_{\ell^p}$$

almost everywhere. Integrating in s, we get

$$\left(\sum_{n=0}^{\infty} |A_n|^p\right)^{\frac{1}{p}} \le \int_0^1 s^{-\frac{1}{p}} \|\{d_j\}\|_{\ell^p} \, ds = p' \|\{d_j\}\|_{\ell^p},$$

as desired. □

It is not difficult to check the optimality of the obtained constant. For any $\delta \in (0, \frac{1}{p'})$, we set

$$d_j = (j+1)^{\frac{1}{p'}-\delta} - j^{\frac{1}{p'}-\delta},$$

with $j = 0, 1, \ldots$. Then any constant $C > 0$ such that

$$\|\{A_n\}\|_{\ell^p} \le C \|\{d_j\}\|_{\ell^p},$$

for every $\{d_j\} \in \ell^p$, must satisfy

$$\sigma(\delta) := \left(\sum_{n=0}^{\infty} \frac{1}{(n+1)^{1+p\delta}}\right)^{\frac{1}{p}}$$

$$\le C\left(1 + \left(\frac{1}{p'} - \delta\right)\right)^p \sum_{n=1}^{\infty} \frac{1}{n^{1+p\delta}}\right)^{\frac{1}{p}}$$

$$\sim \frac{C}{p'} \sigma(\delta),$$

as $\delta \to 0+$, since $\sigma(\delta) \to \infty$. Therefore, $C \ge p'$.

7.2 Hardy Inequalities for Hausdorff Operators

The idea of this section is multi-purpose. Of course, it is one more illustration of how the basic results and methods work. But also it is a kind of supplement to that basic matter. Hardy's inequalities have already been mentioned and illustrated. Here we give them in a very developed form. The proof of the generalization also gives a clear idea how to treat the usual case if one does not wish to deal with extensions.

We are motivated by the fact that the Hardy and Bellman operators are (simple) partial cases of much more general Hausdorff operators defined, by means of a kernel function φ, in most works as

$$(Kf)(x) = (K_\varphi f)(x) = \int_{\mathbb{R}_+} \frac{\varphi(t)}{t} f\left(\frac{x}{t}\right) dt. \tag{7.13}$$

In the paper [159], this operator in full generality had for the first time been studied on the real Hardy space H^1 rather than on Lebesgue spaces. This gave an impact to a series of works on a variety of spaces and in arbitrary dimension; the state of affairs in this topic is given in two survey papers [148] and [86]. This is only one of the ways to define Hausdorff operators, for instance, they can be defined on $\mathbb{R}$ or the distortion of the variable in f can be different. This form is convenient because of its symmetry, that is, the right-hand side of (7.13) is equal to

$$\int_{\mathbb{R}_+} \frac{\varphi\left(\frac{x}{t}\right)}{t} f(t)\, dt. \tag{7.14}$$

To make it even closer to the Hardy inequalities business, we will deal with the following simplified form of Hausdorff operators

$$(Kf)(x) = (K_\varphi f)(x) = \int_{\mathbb{R}_+} \varphi\left(\frac{x}{t}\right) f(t)\, dt. \tag{7.15}$$

By this, denoting, as in Sect. 2.1.3 of Chap. 2, by χ_E the indicator function of the set E, we have $Kf(x) = P_x f$ provided $\varphi(s) = \chi_{[1,\infty)}(s)$ and $Kf(x) = Q_x f$ provided $\varphi(s) = \chi_{(0,1]}(s)$.

Our goal is to generalize (7.2) and (7.3) to a wider family of Hausdorff operators. One cannot say that certain attempts of such type have never been made. For example, the operator in [199] may be considered as a Hausdorff operator with a compactly supported kernel. For the general theory of Hardy's inequalities, besides several books, one of which has already been cited [42], the recent survey [112] can be recommended.

Our strategy for extending the part "if" will be to repeat calculations in [81] but then to stop at a proper moment. In other words, the first part of these calculations in [81] is very general while the second one is specific for P_x and Q_x. Definitely, replacing the general K_φ by one of them and continuing the calculations like in [81] will give these two important partial cases.

The necessity part is more specific and assumes additional properties of the Hausdorff operator that automatically hold for the Hardy operator or for the Bellman operator.

In fact, "playing" with φ, one can derive a variety of Hardy type sufficient conditions. The question is which one is sharp, that is, becomes necessary for the whole class of functions $f \in L_v^p$. We give one of them, considering also the problem whether the $K_\varphi f$ is well-defined for $f \in L_v^p$. As above, the constants in the defining conditions (7.18), (7.19) and (7.20) are independent of the varying parameter, t in this case.

Theorem 7.16 Let $1 < p \le q < \infty$, $f \in L_v^p$, $\varphi \in L^\infty$, and $\varphi, f \ge 0$. We have

$$\left(\int_0^\infty u(x)^q \left|\int_0^\infty \varphi\left(\frac{x}{t}\right) f(t)\, dt\right|^q dx\right)^{\frac{1}{q}} \lesssim \left(\int_0^\infty f(t)^p v(t)^p dt\right)^{\frac{1}{p}} \tag{7.17}$$

provided

$$\int_0^\infty u(x)^q \varphi(\frac{x}{t})^{\frac{q}{p}} \left(\int_0^t \varphi(\frac{x}{s}) v(s)^{-p'} ds \right)^{\frac{q}{pp'}}$$

$$\times \left(\int_0^\infty \varphi(\frac{x}{s}) v(s)^{-p'} ds \right)^{\frac{q}{p'p'}} dx \lesssim 1. \qquad (7.18)$$

If

$$\varphi(\frac{x}{t}) \int_0^t v(s)^{-p'} ds \lesssim \int_0^t \varphi(\frac{x}{s}) v(s)^{-p'} ds, \qquad (7.19)$$

then

$$\int_0^\infty u(x)^q \varphi(\frac{x}{t})^{\frac{q}{p}} \left(\int_0^t \varphi(\frac{x}{s}) v(s)^{-p'} ds \right)^{\frac{q}{p'}} dx \lesssim 1 \qquad (7.20)$$

is necessary for (7.17) to hold.

Proof Let us first check that the Hausdorff operator is well-defined. Indeed, by Hölder's inequality we get

$$\int_0^\infty \varphi(\frac{x}{t}) f(t) dt \leq \left(\int_0^\infty \varphi(\frac{x}{t}) f(t)^p v(t)^p dt \right)^{\frac{1}{p}} \left(\int_0^\infty \varphi(\frac{x}{t}) v(t)^{-p'} dt \right)^{\frac{1}{p'}}.$$

The first integral on the right is finite since $f \in L_v^p$ and $\varphi \in L^\infty$, while the second one is finite since (7.18) assumes it finiteness almost everywhere.

Let now (7.18) hold. By Hölder's inequality, we obtain

$$\int_0^\infty u(x)^q \left| \int_0^\infty \varphi(\frac{x}{t}) f(t) dt \right|^q dx$$

$$= \int_0^\infty u(x)^q \left| \int_0^\infty \varphi(\frac{x}{t})^{\frac{1}{p}} f(t) v(t) h(x,t) \varphi(\frac{x}{t})^{\frac{1}{p'}} v(t)^{-1} h(x,t)^{-1} dt \right|^q dx$$

$$\leq \int_0^\infty u(x)^q \left(\int_0^\infty \varphi(\frac{x}{t}) f(t)^p v(t)^p h(x,t)^p dt \right)^{\frac{q}{p}}$$

$$\times \left(\int_0^\infty \varphi(\frac{x}{t}) v(t)^{-p'} h(x,t)^{-p'} dt \right)^{\frac{q}{p'}} dx, \qquad (7.21)$$

where

$$h(x,t) = \left(\int_0^t \varphi(\frac{x}{s}) v(s)^{-p'} ds \right)^{\frac{1}{pp'}}.$$

The last integral on the right can be reduced, by substituting $y = h(x, t)^{pp'}$, to

$$\int_0^\infty \varphi(\frac{x}{t})^{-p'} v(t)^{-p'} h(x, t)^{-p'} \, dt = \int_{h(x,0)^{pp'}}^{h(x,\infty)^{pp'}} z^{-\frac{1}{p}} \, dz$$

$$= p' \left(\int_0^\infty \varphi(\frac{x}{s}) v(s)^{-p'} \, ds \right)^{\frac{1}{p'}}.$$

Applying the generalized Minkowski inequality to the right-hand side of (7.21), which is

$$\int_0^\infty \left(u(x)^p \int_0^\infty \varphi(\frac{x}{t}) f(t)^p v(t)^p h(x, t)^p \, dt \left(\int_0^\infty \varphi(\frac{x}{s}) v(s)^{-p'} \, ds \right)^{\frac{p}{p'p'}} \right)^{\frac{q}{p}} dx$$

times $(p')^{q/p'}$, we get

$$\left[\int_0^\infty f(t)^p v(t)^p \left(\int_0^\infty u(x)^q \varphi(\frac{x}{t})^{\frac{q}{p}} h(x, t)^q h(x, \infty)^{\frac{qp}{p'}} dx \right)^{\frac{p}{q}} dt \right]^{\frac{q}{p}}$$

times the same constant multiple. By (7.18), it is dominated by the right-hand side of (7.17) in the power q, as desired.

Conversely, assume that (7.17) holds. This means that it is true also in the form

$$\left(\int_0^\infty u(x)^q \left| \int_0^t \varphi(\frac{x}{s}) f(s) \, ds \right|^q dx \right)^{\frac{1}{q}} \lesssim \left(\int_0^t f(s)^p v(s)^p \, ds \right)^{\frac{1}{p}}$$

for each t. For $f(s) = v(s)^{-p'}$, we reduce it to

$$\left(\int_0^\infty u(x)^q \varphi(\frac{x}{t})^{\frac{q}{p}} \left| \int_0^t \varphi(\frac{x}{s}) v(s)^{-p'} \, dt \right|^q \left(\varphi(\frac{x}{t}) \int_0^t v(s)^{-p'} \, ds \right)^{-\frac{q}{p}} dx \right)^{\frac{1}{q}} \lesssim 1.$$

Applying now (7.19), we arrive at (7.20). The proof is complete.   □

Remark 7.22   As mentioned, taking $\varphi(s) = \chi_{(1,\infty)}(s)$, we reduce (7.20) to (7.5). Condition (7.19) is satisfied in this case automatically. However, (7.18) also follows from (7.5) in this case. This is just the second part of the proof of the sufficiency in [81]. Indeed, for the chosen φ, condition (7.18) is

$$\int_t^\infty u(x)^q \left(\int_0^t v(s)^{-p'} \, ds \right)^{\frac{q}{pp'}} \left(\int_0^x v(s)^{-p'} \, ds \right)^{\frac{q}{p'p'}} dx \lesssim 1.$$

Applying (7.5), we estimate the left-hand side by

$$\left(\int_0^t v(s)^{-p'} ds\right)^{\frac{q}{pp'}} \int_t^\infty u(x)^q \left(\int_x^\infty u(s)^q ds\right)^{-\frac{1}{p'}} dx.$$

After integration in x, it becomes

$$\left(\int_0^t v(s)^{-p'} ds\right)^{\frac{q}{pp'}} p \left(\int_t^\infty u(s)^q ds\right)^{-\frac{1}{p}}.$$

Assuming this value to be uniformly bounded gives exactly (7.5).

For a version of Hardy's inequality in a somewhat different notation, see (8.54) in the following chapter.

Counterparts of the obtained results are related to defining $h(x, t)$ by integrals over (t, ∞).

Theorem 7.23 *Let* $1 < p \leq q < \infty$, $f \in L_v^p$, $\varphi \in L^\infty$, *and* $\varphi, f \geq 0$. *We have* (7.17) *provided*

$$\int_0^\infty u(x)^q \varphi(\frac{x}{t})^{\frac{q}{p}} \left(\int_t^\infty \varphi(\frac{x}{s}) v(s)^{-p'} ds\right)^{\frac{q}{pp'}}$$

$$\times \left(\int_0^\infty \varphi(\frac{x}{s}) v(s)^{-p'} ds\right)^{\frac{q}{p'p'}} dx \lesssim 1. \tag{7.24}$$

If

$$\varphi(\frac{x}{t}) \int_t^\infty v(s)^{-p'} ds \lesssim \int_t^\infty \varphi(\frac{x}{s}) v(s)^{-p'} ds, \tag{7.25}$$

then

$$\int_0^\infty u(x)^q \varphi(\frac{x}{t})^{\frac{q}{p}} \left(\int_t^\infty \varphi(\frac{x}{s}) v(s)^{-p'} ds\right)^{\frac{q}{p'}} dx \lesssim 1 \tag{7.26}$$

is necessary for (7.17) *to hold.*

Remark 7.27 Similarly to the above, taking $\varphi(s) = \chi_{(0,1)}(s)$, we reduce (7.26) to (7.6). Condition (7.25) is satisfied in this case automatically. However, (7.24) also follows from (7.6) in this case.

Remark 7.28 Condition (7.18) can be made necessary if we assume

$$\varphi(\frac{x}{t}) \int_0^t v(s)^{-p'} ds$$

$$\lesssim \left(\int_0^t \varphi(\frac{x}{s}) v(s)^{-p'} ds \right)^{p(1-\frac{1}{pp'})} \left(\int_0^\infty \varphi(\frac{x}{s}) v(s)^{-p'} ds \right)^{-\frac{p}{p'p'}}$$

instead of (7.19). However, the class of Hausdorff operators for which this is true does not seem to be rich enough, since it does not include the classical Hardy operator. A similar remark can be said about (7.24).

These results are only a part of those obtained in [155]. For example, similar results are given for probably the most general one-dimensional operator of the form apparently first considered in [136] (see also [137]). Given an odd function a such that $|a(t)|$ is decreasing, positive, and bijective on $(0, \infty)$ (so that both $|a|$ and $\frac{1}{|a|}$ possess inverse functions in such an interval), we define

$$(\mathcal{K}f)(x) = (\mathcal{K}_{\varphi,a}f)(x) = \int_{\mathbb{R}} \varphi(t) |a(t)| f(a(t)x) dt . \qquad (7.29)$$

It is clear that (7.13) corresponds to (7.29) with $a(t) = \frac{1}{t}$. However, even for this very general form the results can be derived from Theorems 7.20 and 7.23 by substitutions in a routine way.

Chapter 8
Certain Applications

In this chapter, we present solutions of certain problems that illustrate how the introduced tools work. As mentioned, we have started with such an approach in the previous chapter, where certain basics of the topic (of two topics, in fact) were given together with applications. We now continue this line. The present book, at least the first five chapters, mirrors, in a sense, the first chapter of the book [34], of course, in an extended and updated manner. That chapter had been written as a toolkit for certain one-dimensional applications. They form the first, one-dimensional part of [34]. However, we are sure that the toolkit presented in the previous chapters is more universal and in general appropriate for a vast number of problems. And if not sufficient to solve a problem completely, it provides many necessary ingredients. Naturally, there are some problems not included in [34] (just because they were not in the strict framework of the book) or even more recent ones. On the other hand, to leave the instructive text without applications could by no means be a good idea. We hope that the following issues illustrate the application points of the above content but also show how the choice of the above matter springs to life. The results below are not massive and cumbersome. In line with the wording of the title of the book, they are examples how the mushrooms collected along the described path can be cooked or fried. The problems considered below also explain, at least to a certain extent, what were the reasons for the choice of the matter in the preceding chapters, or, again in terms of the title of the book, why just this specific path had been chosen. One more goal of this chapter is to consider deeper, on concrete problems, certain topics only touched on in the previous basic part.

8.1 Interpolation Properties of a Scale of Spaces

The notion of interpolation has already been mentioned, even along with hints at its applications. Here we get to know this machinery better. The spaces O_p could be a good model for the so-called *real method* of interpolation. However, following

E. Liflyand, *Harmonic Analysis on the Real Line*, Pathways in Mathematics,
https://doi.org/10.1007/978-3-030-81892-0_8

[140], we extend this consideration to the spaces $O_{p,r}$ endowed with the norms

$$\|f\|_{O_{p,r}} = \left(\int_0^\infty \left[\frac{1}{u} \int_{u \leq |\xi| \leq 2u} |f(\xi)|^p d\xi \right]^{\frac{r}{p}} du \right)^{\frac{1}{r}} \quad (1 \leq p, r < \infty),$$

$$\|f\|_{O_{\infty,r}} = \left(\int_0^\infty \left[\sup_{u \leq |\xi| \leq 2u} |f(\xi)| \right]^r du \right)^{\frac{1}{r}} \quad (1 \leq r < \infty),$$

$$\|f\|_{O_{p,\infty}} = \sup_{u>0} \left(\frac{1}{u} \int_{u \leq |\xi| \leq 2u} |f(\xi)|^p d\xi \right)^{\frac{1}{p}} \quad (1 \leq p < \infty).$$

The case $p = r = \infty$ is naturally reduced to $O_{\infty,\infty} = L^\infty$; moreover, for any $1 \leq p < \infty$, we have $O_{p,p} = L^p$. In the case $r = 1$, we arrive exactly at $O_{p,1} = O_p$, which has been considered and discussed above in Sect. 6.4 of Chap. 6. In [142], interpolation properties of O_p as well as $O_{p,r}$ were studied by means of L^p interpolation properties of vector-valued functions (see, e.g., [7, Theorem 5.1.2.]). Here we study the same object by means of K-functionals, that is, by the *real interpolation method*.

To formulate the results, we recall that the interpolation space $(X_1, X_2)_{\theta,p}$ of the couple X_1 and X_2 consists of all $f \in X_1 + X_2$ such that

$$\|f\|_{\theta,p} = \left(\int_0^\infty [t^{-\theta} K(f, t; X_1, X_2)]^p \frac{dt}{t} \right)^{\frac{1}{p}} < \infty,$$

where

$$0 < \theta < 1, \quad 1 \leq p < \infty,$$

and

$$K(f, t; X_1, X_2) = \inf_{f=g+h} (\|g\|_{X_1} + t\|h\|_{X_2})$$

is the K-functional of the two spaces. It is well-known that for $0 < \theta < 1, 1 \leq p_0 < p_1 \leq \infty$, and

$$\frac{1}{p} = \frac{1-\theta}{p_0} + \frac{\theta}{p_1},$$

we have

$$(L^{p_0}, L^{p_1})_{\theta,p} = L^p. \tag{8.1}$$

8.1.1 Results

One may expect that the spaces $O_p = O_{p,1}$, which seem to be the most important among the considered ones, are interpolated in the same way as in (8.1). It turns out that this is not the case for them; moreover also for most of $O_{p,r}$ which are of the vector-valued type. The (8.1) type interpolation occurs when the first parameter is fixed (see Theorem 8.4); however, to date this case seems to be of less importance. In the case where the second parameter is fixed while the first one varies, the interpolation space is of a special structure and contains an "irregular" part in regard to $O_{p,r}$, with $p \neq r$.

Recall also that the non-increasing rearrangement of a measurable function f is defined in Sect. 2.1.2 of Chap. 2 as

$$f^*(t) = \sup_{|E|=t} \inf_{x \in E} |f(x)| \quad (0 < t < \infty).$$

Given $p_0 < p_1$, we denote

$$p^* = \frac{p_0 p_1}{p_1 - p_0}.$$

By I_j, we denote the (doubled) dyadic interval

$$I_j = \{\xi : 2^j \leq |\xi| \leq 2^{j+1}\}.$$

Let $f_j = f \cdot \chi_{I_j}$, where χ_E denotes, as everywhere, the indicator function of the set E.

Theorem 8.2 *Let $0 < \theta < 1$, $1 \leq p_0 < p_1 \leq \infty$, and $\frac{1}{p} = \frac{1-\theta}{p_0} + \frac{\theta}{p_1}$. Then*

(i) *The following equivalence holds for all $r \geq 1$:*

$$K(f, t; O_{p_0,r}, O_{p_1,r}) \asymp \left(\sum_{j \in \mathbb{Z}} 2^{j(1-\frac{r}{p_0})} \left(\int_0^{2^j t^{p^*}} [f_j^*(\tau)]^{p_0} d\tau \right)^{\frac{r}{p_0}} \right)^{\frac{1}{r}} \tag{8.3}$$

$$+ t \left(\sum_{j \in \mathbb{Z}} 2^{j(1-\frac{r}{p_1})} \left(\int_{2^j t^{p^*}}^{\infty} [f_j^*(\tau)]^{p_1} d\tau \right)^{\frac{r}{p_1}} \right)^{\frac{1}{r}} \quad (0 < t < \infty).$$

(ii) *If $1 \leq r < p$, then*

$$O_{p,r} \subset (O_{p_0,r}, O_{p_1,r})_{\theta,p},$$

and the converse embedding is invalid: there exists a function

$$f \in (O_{p_0,r}, O_{p_1,r})_{\theta,p} \setminus O_{p,r}.$$

(iii) If $r = p$, then

$$(O_{p_0,p}, O_{p_1,p})_{\theta,p} = O_{p,p} = L^p.$$

(iv) If $p < r < \infty$, then

$$(O_{p_0,r}, O_{p_1,r})_{\theta,p} \subset O_{p,r},$$

and the converse embedding is invalid: there exists a function

$$f \in O_{p,r} \setminus (O_{p_0,r}, O_{p_1,r})_{\theta,p}.$$

Theorem 8.4 *Let $1 \le r_0 < r_1 \le \infty$, $0 < \theta < 1$, and $\frac{1}{r} = \frac{1-\theta}{r_0} + \frac{\theta}{r_1}$. Then for all $1 \le p \le \infty$,*

$$(O_{p,r_0}, O_{p,r_1})_{\theta,r} = O_{p,r}. \tag{8.5}$$

8.1.2   Proofs

We will now proceed to the proofs of these two theorems.

Proof of Theorem 8.2 First, we note that the norm in $O_{p,r}$ can be represented in a usual way via doubled dyadic intervals $I_j = \{\xi : 2^j \le |\xi| \le 2^{j+1}\}$:

$$\|f\|_{O_{p,r}} \asymp \left(\sum_{j \in \mathbb{Z}} 2^{j(1-\frac{r}{p})} \left(\int_{I_j} |f(\xi)|^p d\xi \right)^{\frac{r}{p}} \right)^{\frac{1}{r}}, \tag{8.6}$$

with $1 \le p \le \infty$ and $1 \le r < \infty$. Now (i) is a simple consequence of (8.6) and the known formula for the K-functional for (L^{p_0}, L^{p_1}) couple (see, e.g., [7, p.124]):

$$K(g, t; L^{p_0}, L^{p_1}) \asymp \left(\int_0^{t^{p*}} [g^*(\tau)]^{p_0} d\tau \right)^{\frac{1}{p_0}} + t \left(\int_{t^{p*}}^{\infty} [g^*(\tau)]^{p_1} d\tau \right)^{\frac{1}{p_1}}.$$

To be precise, a misprint occured in [7], namely, t is omitted before the last term on the right; in [6, p.308] the needed result is given in a more general form.

It remains to observe that for $t > 2^{\frac{1}{p*}}$, the right side of (8.3) is equivalent to $\|f\|_{O_{p_0,r}}$.

We now prove (ii). Using (8.3) and Minkowski's inequality, we obtain

$$\|f\|_{(O_{p_0,r},O_{p_1,r})_{\theta,p}}$$

$$\leq C\left(\sum_{j\in\mathbb{Z}}2^{j(1-r/p_0)}\left(\int_0^\infty t^{-\theta p}\left(\int_0^{2^j t^{p^*}}[f_j^*(\tau)]^{p_0}d\tau\right)^{\frac{p}{p_0}}\frac{dt}{t}\right)^{\frac{r}{p}}\right)^{\frac{1}{r}}$$

$$+ C\left(\sum_{j\in\mathbb{Z}}2^{j(1-\frac{r}{p_1})}\left(\int_0^\infty t^{(1-\theta)p}\left(\int_{2^j t^{p^*}}^\infty[f_j^*(\tau)]^{p_1}d\tau\right)^{\frac{p}{p_1}}\frac{dt}{t}\right)^{\frac{r}{p}}\right)^{\frac{1}{r}}. \tag{8.7}$$

The substitution $t^{p^*}\to t$ and Hardy's inequality (see, e.g., [6, p.124], Sect. 5.5 of Chap. 5 and also Chap. 7) show that the first sum in (8.7) is at most (maybe times an absolute constant multiple)

$$\left(\sum_{j\in\mathbb{Z}}2^{j(1-\frac{r}{p_0})}\left(\int_0^\infty\left(\frac{1}{t}\int_0^{2^j t}[f_j^*(\tau)]^{p_0}d\tau\right)^{\frac{p}{p_0}}dt\right)^{\frac{r}{p}}\right)^{\frac{1}{r}}$$

$$\leq C\left(\sum_{j\in\mathbb{Z}}2^{j(1-\frac{r}{p})}\left(\int_{I_j}|f(\xi)|^p d\xi\right)^{\frac{r}{p}}\right)^{\frac{1}{r}}\leq C\|f\|_{O_{p,r}}.$$

Since the rearrangement is a non-increasing function, it follows from the same substitution that the second sum in (8.7) is bounded by

$$C\left(\sum_{j\in\mathbb{Z}}2^{j(1-\frac{r}{p_1})}\left(\int_0^\infty\left(\frac{1}{t}\int_{2^j t}^\infty[f_j^*(\tau)]^{p_1}d\tau\right)^{\frac{p}{p_1}}dt\right)^{\frac{r}{p}}\right)^{\frac{1}{r}}$$

$$\leq C\left(\sum_{j\in\mathbb{Z}}2^{j(1-\frac{r}{p})}\left(\int_0^\infty\left(\sum_{k=0}^\infty 2^k[f_j^*(2^k t)]^{p_1}\right)^{\frac{p}{p_1}}dt\right)^{\frac{r}{p}}\right)^{\frac{1}{r}}$$

$$\leq C\left(\sum_{j\in\mathbb{Z}}2^{j(1-\frac{r}{p})}\left(\sum_{k=0}^\infty 2^{\frac{kp}{p_1}}\int_0^\infty[f_j^*(2^k t)]^p dt\right)^{\frac{r}{p}}\right)^{\frac{1}{r}}$$

$$= C\left(\sum_{j\in\mathbb{Z}}2^{j(1-\frac{r}{p})}\left(\sum_{k=0}^\infty\frac{1}{2^{k(1-\frac{p}{p_1})}}\int_{I_j}|f(\xi)|^p d\xi\right)^{\frac{r}{p}}\right)^{\frac{1}{r}}\leq C\|f\|_{O_{p,r}}.$$

Therefore, we get

$$\|f\|_{(O_{p_0,r},O_{p_1,r})_{\theta,p}}\leq C\|f\|_{O_{p,r}}, \tag{8.8}$$

which proves the embedding $O_{p,r}\subset(O_{p_0,r},O_{p_1,r})_{\theta,p}$.

To complete the second item, let us show that there exists a function $f \in (O_{p_0,r}, O_{p_1,r})_{\theta,p}$ such that $f \notin O_{p,r}$. Let f be of the form

$$f = \sum_{j=1}^{\infty} c_j \chi_{(2^j, 2^j(1+a_j))},$$

where sequences $\{c_j\}$, $\{a_j\}$ will be specified later on. We now only point out that $a_j < 1$, $\{a_j\}$ is decreasing and tends to zero. It is easy to see that for this function f, we have

$$\|f\|_{O_{s,r}} \asymp \left(\sum_{j=1}^{\infty} 2^j c_j^r a_j^{\frac{s}{s}} \right)^{\frac{1}{r}}. \tag{8.9}$$

Next, a simple calculation shows that

$$\left(\int_0^{2^j t^{p^*}} [f_j^*(\tau)]^{p_0} d\tau \right)^{\frac{r}{p_0}} = \begin{cases} (2^j)^{\frac{r}{p_0}} c_j^r t^{p^* \frac{r}{p_0}}, & 0 < t \le a_j^{\frac{1}{p^*}}, \\[2mm] c_j^r (2^j a_j)^{\frac{r}{p_0}}, & t > a_j^{\frac{1}{p^*}}, \end{cases}$$

and

$$\left(\int_{2^j t^{p^*}}^{\infty} [f_j^*(\tau)]_1^p d\tau \right)^{\frac{r}{p_1}} \le c_j^r (2^j a_j)^{\frac{r}{p_1}} \chi_{(0, a_j^{\frac{1}{p^*}})}(t).$$

Therefore,

$$K(f, t; O_{p_0,r}, O_{p_1,r})$$

$$\asymp \left(\sum_{j=1}^{\infty} \left((2^j c_j^r t^{p^* \frac{r}{p_0}} + 2^j c_j^r a_j^{\frac{r}{p_1}} t^r) \chi_{(0, a_j^{\frac{1}{p^*}})}(t) + 2^j c_j^r a_j^{\frac{r}{p_0}} \chi_{(a_j^{\frac{1}{p^*}}, \infty)}(t) \right) \right)^{\frac{1}{r}}$$

$$\asymp \left(\sum_{k=1}^{\infty} \left(t^{p^* \frac{r}{p_0}} \sum_{j=1}^{k} 2^j c_j^r + t^r \sum_{j=1}^{k} 2^j c_j^r a_j^{r/p_1} + \sum_{j=k+1}^{\infty} 2^j c_j^r a_j^{\frac{r}{p_0}} \right) \chi_{(a_{k+1}^{\frac{1}{p^*}}, a_k^{\frac{1}{p^*}})}(t) \right.$$

$$\left. + \|f\|_{O_{p_0,r}} \chi_{(a_1^{\frac{1}{p^*}}, \infty)}(t) \right)^{\frac{1}{r}}.$$

From this we get

$$\|f\|^p_{(O_{p_0,r},O_{p_1,r})_{\theta,p}}$$

$$\leq C\sum_{k=1}^{\infty}\left\{\left(\sum_{j=1}^{k}2^jc_j^r\right)^{\frac{p}{r}}\int_{a_{k+1}^{\frac{1}{p^*}}}^{a_k^{\frac{1}{p^*}}}t^{p^*-1}\,dt+\left(\sum_{j=1}^{k}2^jc_j^ra_j^{\frac{r}{p_1}}\right)^{\frac{p}{r}}\int_{a_{k+1}^{\frac{1}{p^*}}}^{a_k^{\frac{1}{p^*}}}t^{\frac{p^*(p_1-p)}{p_1}-1}\,dt\right.$$

$$+\left(\sum_{j=k+1}^{\infty}2^jc_j^ra_j^{\frac{r}{p_0}}\right)^{\frac{p}{r}}\int_{a_{k+1}^{\frac{1}{p^*}}}^{a_k^{\frac{1}{p^*}}}\frac{dt}{t^{\frac{p^*(p-p_0)}{p_0}+1}}\left.\right\}+C\|f\|^{\frac{p}{r}}_{O_{p_0,r}}$$

$$\leq C\sum_{k=1}^{\infty}\left\{\left(\sum_{j=1}^{k}2^jc_j^r\right)^{\frac{p}{r}}(a_k-a_{k+1})+\left(\sum_{j=1}^{k}2^jc_j^ra_j^{r/p_1}\right)^{\frac{p}{r}}\left(a_k^{\frac{p_1-p}{p_1}}-a_{k+1}^{\frac{p_1-p}{p_1}}\right)\right.$$

$$+\left(\sum_{j=k+1}^{\infty}2^jc_j^ra_j^{\frac{r}{p_0}}\right)^{\frac{p}{r}}\left(\frac{1}{a_{k+1}}\right)^{\frac{p-p_0}{p_0}+\frac{1}{p^*}}\left(a_k^{\frac{1}{p^*}}-a_{k+1}^{\frac{1}{p^*}}\right)\left.\right\}$$

$$+C\|f\|^{\frac{p}{r}}_{O_{p_0,r}}.\tag{8.10}$$

We are now in a position to specify the sequences $\{c_j\}$, $\{a_j\}$ so that the right-hand side of (8.10) be finite, while the series in (8.9) be divergent for $s=p$. We first set

$$c_j=2^{-\frac{j}{r}},\quad a_j=(j\log(j+1))^{-\frac{p}{r}}.$$

Then, in view of (8.9),

$$\|f\|^r_{O_{p_0,r}}\asymp\sum_{j=1}^{\infty}\frac{1}{(j\log(j+1))^{\frac{p}{p_0}}}<\infty,$$

while

$$\|f\|^r_{O_{p,r}}\asymp\sum_{j=1}^{\infty}\frac{1}{j\log(j+1)}=\infty.$$

It remains to show that the series on the right-hand side of (8.10) converges. Simple estimates yield

$$a_k-a_{k+1}\leq\frac{C}{k^{\frac{p}{r}+1}(\log(k+1))^{\frac{p}{r}}},$$

$$a_k^{\frac{p_1-p}{p_1}}-a_{k+1}^{\frac{p_1-p}{p_1}}\leq\frac{C}{k^{\frac{p(p_1-p)}{rp_1}+1}(\log(k+1))^{\frac{p(p_1-p)}{rp_1}}},$$

$$\left(\frac{1}{a_{k+1}}\right)^{\frac{p-p_0}{p_0}+\frac{1}{p^*}}\left(a_k^{\frac{1}{p^*}}-a_{k+1}^{\frac{1}{p^*}}\right)\leq Ck^{\frac{p(p-p_0)}{rp_0}-1}(\log(k+1))^{\frac{p(p-p_0)}{rp_0}},$$

and

$$\left(\sum_{j=1}^{k} 2^j c_j^r\right)^{\frac{p}{r}} = k^{\frac{p}{r}},$$

$$\left(\sum_{j=1}^{k} 2^j c_j^r a_j^{\frac{r}{p_1}}\right)^{\frac{p}{r}} \le C \frac{k^{\frac{p(p_1-p)}{rp_1}}}{(\log(k+1))^{\frac{p^2}{rp_1}}},$$

$$\left(\sum_{j=k+1}^{\infty} 2^j c_j^r a_j^{\frac{r}{p_0}}\right)^{\frac{p}{r}} \le \frac{C}{k^{\frac{p(p-p_0)}{rp_0}} (\log(k+1))^{\frac{p^2}{rp_0}}}.$$

Hence, the series on the right-hand side of (8.10) is at most

$$\sum_{k=1}^{\infty} \frac{1}{k(\log(k+1))^{\frac{p}{r}}} < \infty,$$

times some absolute constant multiple, as required.

To prove (iii), it suffices to demonstrate that the estimate converse to (8.8) holds. Indeed, in this case (8.3) and change of variables $2^j t^{p^*} \to t$ yield

$$\|f\|_{(O_{p_0,p}, O_{p_1,p})_{\theta,p}} \asymp C \left(\sum_{j\in\mathbb{Z}} 2^{j(1-\frac{p}{p_0})} \int_0^{\infty} t^{-\theta p} \left(\int_0^{2^j t^{p^*}} [f_j^*(\tau)]^{p_0} d\tau\right)^{\frac{p}{p_0}} \frac{dt}{t}\right)^{\frac{1}{p}}$$

$$+ C \left(\sum_{j\in\mathbb{Z}} 2^{j(1-\frac{p}{p_1})} \int_0^{\infty} t^{(1-\theta)p} \left(\int_{2^j t^{p^*}}^{\infty} [f_j^*(\tau)]^{p_1} d\tau\right)^{\frac{p}{p_1}} \frac{dt}{t}\right)^{\frac{1}{p}}$$

$$\asymp C \left(\sum_{j\in\mathbb{Z}} \int_0^{\infty} \left(\frac{1}{t} \int_0^t [f_j^*(\tau)]^{p_0} d\tau\right)^{\frac{p}{p_0}} dt\right)^{\frac{1}{p}}$$

$$+ C \left(\sum_{j\in\mathbb{Z}} \int_0^{\infty} \left(\frac{1}{t} \int_t^{\infty} [f_j^*(\tau)]^{p_1} d\tau\right)^{\frac{p}{p_1}} dt\right)^{\frac{1}{p}}$$

$$\ge C \left(\sum_{j\in\mathbb{Z}} \int_0^{\infty} [f_j^*(t)]^p dt\right)^{\frac{1}{p}} + C \left(\sum_{j\in\mathbb{Z}} \int_0^{\infty} [f_j^*(2t)]^p dt\right)^{\frac{1}{p}}$$

$$\ge C\|f\|_{L^p}, \tag{8.11}$$

and we are done.

We now consider the last case $p < r < \infty$. Applying (8.3) and Hölder's inequality yields

$$\|f\|_{O_{p,r}} \leq C\left(\sum_{j\in\mathbb{Z}} 2^j \left(\int_0^2 [f_j^*(2^j t)]^p dt\right)^{\frac{r}{p}}\right)^{\frac{1}{r}}$$

$$\leq C\left(\sum_{j\in\mathbb{Z}} 2^{\frac{jp}{r}} \int_0^2 [f_j^*(2^j t)]^p dt\right)^{\frac{1}{p}}$$

$$\leq C\left(\int_0^2 \left(\sum_{j\in\mathbb{Z}} 2^j [f_j^*(2^j t)]^r\right)^{\frac{p}{r}} dt\right)^{\frac{1}{p}}$$

$$\leq C\left(\int_0^2 \left(\sum_{j\in\mathbb{Z}} 2^{j(1-\frac{r}{p_0})}\left(\frac{1}{t}\int_0^{2^j t} [f_j^*(\tau)]^{p_0} d\tau\right)^{\frac{r}{p_0}}\right)^{\frac{p}{r}} dt\right)^{\frac{1}{p}}$$

$$\leq C\left(\int_0^\infty t^{-\theta p}\left(\sum_{j\in\mathbb{Z}} 2^{j(1-\frac{r}{p_0})}\left(\int_0^{2^j t^{p^*}} [f_j^*(\tau)]^{p_0} d\tau\right)^{\frac{r}{p_0}}\right)^{\frac{p}{r}} \frac{dt}{t}\right)^{\frac{1}{p}}$$

$$\leq C\|f\|_{(O_{p_0,r},O_{p_1,r})_{\theta,p}}, \tag{8.12}$$

which gives the embedding

$$(O_{p_0,r}, O_{p_1,r})_{\theta,p} \subset O_{p,r}.$$

To complete the proof, we show that there exists a function

$$f \in O_{p,r} \setminus (O_{p_0,r}, O_{p_1,r})_{\theta,p}.$$

As above, let f be of the form

$$f = \sum_{j=1}^\infty c_j \chi_{(2^j, 2^j(1+a_j))}.$$

It is clear that $(f_j)^*(2^j t) = c_j \chi_{(0,a_j)}(t)$, and hence,

$$\|f\|_{(O_{p_0,r}, A_{p_1,r})_{\theta,p}} \geq C\left(\int_0^2 \left(\sum_{j\in\mathbb{Z}} 2^j [f_j^*(2^j t)]^r\right)^{\frac{p}{r}} dt\right)^{\frac{1}{p}}$$

$$\geq C\sum_{k=1}^\infty \int_{a_{k+1}}^{a_k} \left(\sum_{j=1}^k 2^j [f_j^*(2^j t)]^r\right)^{\frac{p}{r}}$$

$$= \sum_{k=1}^\infty \left(\sum_{j=1}^k 2^j c_j r\right)^{\frac{p}{r}} (a_k - a_{k+1}).$$

Setting now

$$c_j = 2^{-\frac{j}{r}}, \quad a_j = (j^{\frac{p}{r}} \log(j+1))^{-1},$$

we obtain, in view of (8.9),

$$\|f\|^r_{O_{p,r}} \asymp \sum_{j=1}^{\infty} \frac{1}{j(\log(j+1))^{\frac{r}{p}}} < \infty,$$

while

$$\sum_{k=1}^{\infty} \left(\sum_{j=1}^{k} 2^j c_j r\right)^{\frac{p}{r}} (a_k - a_{k+1}) \asymp \sum_{j=1}^{\infty} \frac{1}{j \log(j+1)} = \infty.$$

The proof of the theorem is complete.   □

Remark 8.13 In the proof of (ii) and (iv), the estimates from above and from below, respectively, were used for the norm in the interpolation space (see (8.7) and what follows, and (8.12)). The counterexamples given above show that it is never possible for the corresponding bounds to be bilateral, except for the case $r = p$ (see (8.11)).

Proof of Theorem 8.4 We first consider the case $r_0 = 1$ and $r_1 = \infty$. Denoting

$$T_p f(u) = \left(\frac{1}{u} \int_{u \leq |\xi| \leq 2u} |f(\xi)|^p d\xi\right)^{\frac{1}{p}},$$

let us show that

$$\int_0^t (T_p f)^*(\tau) d\tau \leq K(f,t; O_{p,1}, O_{p,\infty}) \leq 25 \int_0^t (T_p f)^*(\tau) d\tau. \qquad (8.14)$$

The left-hand side of (8.14) is trivial. To prove the right-hand side of (8.14), we set

$$g(x) = f \cdot \chi_{(-t,t)}(x), \quad h(x) = f(x) - g(x).$$

Then

$$\|g\|_{O_{p,1}} = \int_0^t T_p f(u) du \leq \int_0^t (T_p f)^*(\tau) d\tau.$$

We now observe that

$$
\begin{aligned}
T_p f(u) &\le \inf_{s \in (\frac{u}{2}, u)} \left(\frac{1}{u} \int_{s \le |\xi| \le 4s} |f(\xi)|^p d\xi \right)^{\frac{1}{p}} \\
&\le \inf_{s \in (\frac{u}{2}, u)} \left(T_p f(s) + 2 T_p f(2s) \right) \\
&\le \left(T_p f(s) + 2 T_p f(2s) \right)^* (\frac{u}{2}) \le (T_p f)^* (\frac{u}{4}) \\
&\quad + 2 (T_p f)^* (\frac{u}{2}) \le 3 (T_p f)^* (\frac{u}{4}).
\end{aligned}
$$

Since $h(x) = 0$ for $x \in (-t, t)$, we obtain

$$
\begin{aligned}
\|h\|_{O_{p,\infty}} &= \sup_{u > 0} \left(\frac{1}{u} \int_{u \le |\xi| \le 2u} |h(\xi)|^p d\xi \right)^{\frac{1}{p}} \\
&= \sup_{u > \frac{t}{2}} T_p f(u) \le 3 (T_p f)^* (\frac{t}{8}),
\end{aligned}
$$

and thus

$$
\begin{aligned}
K(f, t; O_{p,1}, O_{p,\infty}) &\le \|g\|_{O_{p,1}} + t \|h\|_{O_{p,\infty}} \\
&\le \int_0^t (T_p f)^*(\tau) d\tau + 3t (T_p f)^* (\frac{t}{8}) \\
&\le \int_0^t (T_p f)^*(\tau) d\tau + 24 \int_0^{\frac{t}{8}} (T_p f)^*(\tau) d\tau \\
&\le 25 \int_0^t (T_p f)^*(\tau) d\tau,
\end{aligned}
$$

as required.

It is clear that (8.14) proves the theorem if $r_0 = 1$ and $r_1 = \infty$. One can then apply the Holmstedt reiteration theorem (see, e.g., [6, p.307]) to describe the K-functional for any of the couples (O_{p,r_0}, O_{p,r_1}) and get (8.5) for all $1 < r_0 < r_1 < \infty$. $\qquad\square$

8.2   Fourier Re-expansions

In the 50's (see, e.g., [125] or in more detail [25, Chapters II and VI]), the following problem in harmonic analysis attracted much attention:

Let $\{a_k\}_{k=0}^\infty$ be the sequence of the Fourier coefficients of the absolutely convergent sine (cosine) Fourier series of a function $f : \mathbb{T} = [-\pi, \pi) \to \mathbb{C}$, that is, $\sum |a_k| < \infty$. Under what conditions on $\{a_k\}$ will the re-expansion of $f(t)$

(or $f(t) - f(0)$, respectively) in the cosine (sine) Fourier series also be absolutely convergent?

The condition obtained in the mentioned references is quite simple and is the same in both cases:

$$\sum_{k=1}^{\infty} |a_k| \ln(k+1) < \infty. \tag{8.15}$$

I learned about this problem while being a student, and all my life was impressed by it but only much later proceeded to a serious study (see [149]) of various versions of this puzzle. This is one of the reasons why it has not been touched on in the chapter on Fourier series. A natural step is the study of a similar problem of the integrability of the re-expansion for Fourier transforms of functions defined on $\mathbb{R}_+ = [0, \infty)$. We give necessary and sufficient conditions in terms of belonging of the sine or cosine Fourier transform to a certain Hardy space. The proof differs from that for Fourier series. The reason is that an attempt to repeat that proof, in fact, the direct recalculation, comes across a relation equivalent to Luzin's conjecture on the almost everywhere convergence of the Fourier series. More precisely, more than one hundred years ago Luzin knew (see also the end of Sect. 3.2 in Chap. 3) that the positive answer is equivalent to the vanishing at infinity of the so-called *modulated Carleson integral*. This finally turned out to be the case for L^2 (and some other classes of) functions. However, long before Carleson's solution, it was known that the almost everywhere convergence fails to hold for all L^1 functions (see a nice discussion of these issues in [11]). Because of these circumstances, a different proof for the problem we discuss had to be found. For this, the ideas of Sect. 5.6 in Chap. 5 are of importance. An intent observation of old proofs of the problem in question shows that similar necessary and sufficient conditions hold true for series as well, while (8.15) is just a sufficient condition for belonging to discrete Hardy type spaces; details can be found in [60]. On account of this, in Theorem 5.44 of Sect. 5.5.2 in Chap. 5, we have given a continuous analog of such a sufficient condition. The difference between the two cases can partially be indicated by the fact that a direct analog of (8.15) is not enough and is accompanied by an additional condition (5.46).

We are now in a position to formulate a non-periodic analog of the problem in question. Let, as above,

$$\widehat{f}_c(x) = \int_0^{\infty} f(t) \cos xt \, dt$$

be the cosine Fourier transform of f and

$$\widehat{f}_s(x) = \int_0^{\infty} f(x) \sin xt \, dt$$

be the sine Fourier transform of f, each understood in a certain sense.

Let

$$\int_0^\infty |\widehat{f_c}(x)|\,dx < \infty, \tag{8.16}$$

and hence we have almost everywhere

$$f(t) = \frac{1}{\pi}\int_0^\infty \widehat{f_c}(x)\cos tx\,dx, \tag{8.17}$$

or, in the inverted case,

$$\int_0^\infty |\widehat{f_s}(x)|\,dx < \infty \tag{8.18}$$

and hence almost everywhere

$$f(t) = \frac{1}{\pi}\int_0^\infty \widehat{f_s}(x)\sin tx\,dx. \tag{8.19}$$

The following natural counterparts of the above questions arise:

1. *Under what (additional) conditions on $\widehat{f_c}$ do we get (8.18),*
   or, alternatively,
2. *under what (additional) conditions on $\widehat{f_s}$ do we get (8.16)?*

To formulate the results, recall that the cancelation property like (5.16) holds automatically for the sine transform. As already mentioned, just as for the cosine transform, we need an analog of (5.50) to be satisfied, in the present notation

$$\int_0^\infty \widehat{f_c}(t)\,dt = 0 \tag{8.20}$$

to ensure (5.16).

Theorem 8.21 *In order that the re-expansion $\widehat{f_s}$ of f with the integrable cosine Fourier transform $\widehat{f_c}$ be integrable, it is necessary and sufficient that its Hilbert transform $\mathcal{H}\widehat{f_c}(x)$ be integrable and (8.20) hold.*

Similarly, in order that the re-expansion $\widehat{f_c}$ of f with the integrable sine Fourier transform $\widehat{f_s}$ be integrable, it is necessary and sufficient that its Hilbert transform $\mathcal{H}\widehat{f_s}(x)$ be integrable.

Proof Let (8.16) hold true. Then we can rewrite

$$\widehat{f_s}(x) = \int_0^\infty \left[\frac{1}{\pi}\int_0^\infty \widehat{f_c}(u)\cos tu\,du\right]\sin xt\,dt. \tag{8.22}$$

If one expects the integrability of $\widehat{f}_s$, then the inner integral should vanish at the origin. This follows from (8.20). The right-hand side of (8.22) can be understood in the $(C, 1)$ sense as

$$\frac{1}{\pi} \lim_{N \to \infty} \int_0^N \left(1 - \frac{t}{N}\right) \int_0^\infty \widehat{f}_c(u) \cos tu \, du \, \sin xt \, dt.$$

In virtue of (8.16) we can change the order of integration:

$$\frac{1}{\pi} \lim_{N \to \infty} \int_0^\infty \widehat{f}_c(u) \int_0^N \left(1 - \frac{t}{N}\right) \cos tu \, \sin xt \, dt \, du$$

$$= \frac{1}{\pi} \lim_{N \to \infty} \int_0^\infty \widehat{f}_c(u) \frac{1}{2} \int_0^N \left(1 - \frac{t}{N}\right) [\sin(u + x)t - \sin(u - x)t] \, dt \, du.$$

We now need the next simple formula

$$\int_0^N \left(1 - \frac{t}{N}\right) \sin At \, dt = \frac{1}{A} - \frac{\sin NA}{NA^2}. \tag{8.23}$$

Applying it yields

$$\widehat{f}_s(x) = \frac{1}{\pi} \lim_{N \to \infty} \int_0^\infty \widehat{f}_c(u) \left[\frac{1}{u + x} - \frac{\sin(u + x)N}{N(u + x)^2}\right] du$$

$$- \frac{1}{\pi} \lim_{N \to \infty} \int_0^\infty \widehat{f}_c(u) \left[\frac{1}{u - x} - \frac{\sin(u - x)N}{N(u - x)^2}\right] du$$

$$= I_1 + I_2. \tag{8.24}$$

Let us continue with I_2. Substituting $u - x = t$, we obtain

$$I_2 = -\frac{1}{\pi} \lim_{N \to \infty} \int_{-x}^\infty \widehat{f}_c(x + t) \left[\frac{1}{t} - \frac{\sin Nt}{Nt^2}\right] dt.$$

For I_1, we first substitute $u = -v$. Thus, we get

$$I_1 = \frac{1}{\pi} \lim_{N \to \infty} \int_{-\infty}^0 \widehat{f}_c(-v) \left[\frac{1}{-v + x} - \frac{\sin(-v + x)N}{N(x - v)^2}\right] dv$$

$$= -\frac{1}{\pi} \lim_{N \to \infty} \int_{-\infty}^0 \widehat{f}_c(-v) \left[\frac{1}{v - x} - \frac{\sin(v - x)N}{N(v - x)^2}\right] dv$$

$$= -\frac{1}{\pi} \lim_{N \to \infty} \int_{-\infty}^0 \widehat{f}_c(v) \left[\frac{1}{v - x} - \frac{\sin(v - x)N}{N(v - x)^2}\right] dv.$$

The last equality follows from the evenness of $\widehat{f_c}$. Substituting $v - x = t$, we obtain

$$I_1 = -\frac{1}{\pi} \lim_{N \to \infty} \int_{-\infty}^{-x} \widehat{f_c}(x + t) \left[\frac{1}{t} - \frac{\sin Nt}{Nt^2} \right] dt.$$

Therefore,

$$\widehat{f_s}(x) = -\frac{1}{\pi} \lim_{N \to \infty} \int_{-\infty}^{\infty} \widehat{f_c}(x + t) \left[\frac{1}{t} - \frac{\sin Nt}{Nt^2} \right] dt.$$

We are now in a position to apply Theorem 5.54, which implies that for almost all x, there holds

$$\widehat{f_s}(x) = \mathcal{H}\widehat{f_c}(x). \tag{8.25}$$

We remark that any integrable function satisfies the assumption of Theorem 5.54 and (5.50) (which is (8.20) in the present situation) is necessary for the integrability of the Hilbert transform. Now, let (8.18) holds true. Then we can rewrite

$$\widehat{f_c}(x) = \int_0^\infty \left[\frac{1}{\pi} \int_0^\infty \widehat{f_s}(u) \sin tu \, du \right] \cos xt \, dt. \tag{8.26}$$

As above, the right-hand side can be understood in the $(C, 1)$ sense as

$$\frac{1}{\pi} \lim_{N \to \infty} \int_0^N \left(1 - \frac{t}{N} \right) \int_0^\infty \widehat{f_s}(u) \sin tu \, du \, \cos xt \, dt.$$

In virtue of (8.18) we can change the order of integration:

$$\frac{1}{\pi} \lim_{N \to \infty} \int_0^\infty \widehat{f_s}(u) \int_0^N \left(1 - \frac{t}{N} \right) \sin tu \, \cos xt \, dt \, du$$

$$= \frac{1}{\pi} \lim_{N \to \infty} \int_0^\infty \widehat{f_s}(u) \frac{1}{2} \int_0^N \left(1 - \frac{t}{N} \right) [\sin(u + x)t + \sin(u - x)t] \, dt \, du.$$

Applying (8.23), we get

$$\widehat{f_s}(x) = \frac{1}{\pi} \lim_{N \to \infty} \int_0^\infty \widehat{f_s}(u) \left[\frac{1}{u + x} - \frac{\sin(u + x)N}{N(u + x)^2} \right] du$$

$$+ \frac{1}{\pi} \lim_{N \to \infty} \int_0^\infty \widehat{f_s}(u) \left[\frac{1}{u - x} - \frac{\sin(u - x)N}{N(u - x)^2} \right] du$$

$$= J_1 + J_2. \tag{8.27}$$

Let us proceed to J_2. Substituting $u - x = t$, we obtain

$$J_2 = \frac{1}{\pi} \lim_{N \to \infty} \int_{-x}^{\infty} \widehat{f_s}(x + t) \left[\frac{1}{t} - \frac{\sin Nt}{Nt^2} \right] dt.$$

Treating J_1 as I_1 above, we get

$$J_1 = -\frac{1}{\pi} \lim_{N \to \infty} \int_{-\infty}^{0} \widehat{f_s}(-v) \left[\frac{1}{v - x} - \frac{\sin(v - x)N}{N(v - x)^2} \right] dv$$

$$= \frac{1}{\pi} \lim_{N \to \infty} \int_{-\infty}^{0} \widehat{f_s}(v) \left[\frac{1}{v - x} - \frac{\sin(v - x)N}{N(v - x)^2} \right] dv.$$

The last equality follows from the oddness of $\widehat{f_s}$. Substituting $v - x = t$, we obtain

$$J_1 = \frac{1}{\pi} \lim_{N \to \infty} \int_{-\infty}^{-x} \widehat{f_s}(x + t) \left[\frac{1}{t} - \frac{\sin Nt}{Nt^2} \right] dt.$$

Therefore,

$$\widehat{f_c}(x) = \frac{1}{\pi} \lim_{N \to \infty} \int_{-\infty}^{\infty} \widehat{f_s}(x + t) \left[\frac{1}{t} - \frac{\sin Nt}{Nt^2} \right] dt.$$

Finally, it follows from Theorem 5.54 that for almost all x, we have

$$\widehat{f_c}(x) = -\mathcal{H}\widehat{f_s}(x). \tag{8.28}$$

This completes the proof.   □

Let us comment on the obtained results. In fact, the proof of Theorem 8.21 shows that more general results than stated are obtained. Indeed, formulas (8.25) and (8.28) are more informative than the assertion of Theorem 8.21. To be precise, such formulas are known, see [27, (5.42) and (5.43)]. However, the situation is much more delicate. These formulas are proved in [27] for square integrable functions by applying the Riemann–Lebesgue lemma in an appropriate place (5.44). But in [27, §6.19] more details are given (see also [11]) and it is shown that the possibility to apply the Riemann–Lebesgue lemma in that argument is equivalent to (Carleson's solution of) Luzin's conjecture. In our L^1 setting this is by no means applicable. And, indeed, our proof is different and rests on a less restrictive Theorem 5.54. This agrees well with what E.M. Dyn'kin wrote in his noted survey on singular integrals [11]: "In fact, the theory of singular integrals is a technical subject where ideas cannot be separated from the techniques."

The assertions of Theorem 8.21 can be reformulated in terms of Hardy spaces: *The belonging of $\widehat{f_c}$ (or $\widehat{f_s}$) to the real Hardy space $H^1(\mathbb{R})$ is the necessary and sufficient condition for the integrability of $\widehat{f_s}$ (or $\widehat{f_c}$, respectively).*

8.3 Absolute Convergence

The possibility to represent a function as an absolutely convergent Fourier integral, that is, to belong to Wiener's algebra $W_0(\mathbb{R})$ (see (4.7) and related material in Chap. 4) has been studied by many mathematicians and is of importance in various problems of analysis. For example, in [121] (see also [48, Chapter 4, 7.4]) this problem has been studied, in connection with multipliers, for the following multiplier function:

$$m(x) = \theta(x)\frac{e^{i|x|^{\alpha}}}{|x|^{\beta}}, \qquad (8.29)$$

where θ is a C^{∞} function on $\mathbb{R}$, which vanishes near zero, and equals 1 outside a bounded set, and $\alpha, \beta > 0$. It is known that

(I) If $\frac{\beta}{\alpha} > \frac{1}{2}$, then $m \in W_0(\mathbb{R})$;

(II) If $\frac{\beta}{\alpha} \le \frac{1}{2}$ and $\alpha \ne 1$, then $m \notin W_0(\mathbb{R})$.

Here we present a recent condition for belonging to $W_0(\mathbb{R})$, obtained in [145] as an affirmative solution of a conjecture posed by Trigub. Shortly there after, multidimensional generalizations appeared in [134] and [135].

To start with and to be able to undertake a comparative analysis, let us unite certain known earlier results closely related to our study in the following theorem. First, it is natural to consider functions $f \in W_0(\mathbb{R})$ that satisfy the condition

(N) Let $f \in C_0(\mathbb{R})$, that is, $f \in C(\mathbb{R})$ and $\lim f(t) = 0$ as $|t| \to \infty$, and let f be *locally absolutely continuous on* $\mathbb{R} \setminus \{0\}$.

The following classical result is widely known.

Theorem A Let f satisfy condition **(N)**, $f \in L^p(\mathbb{R})$ with $1 \le p \le 2$, and $f' \in L^q(\mathbb{R})$ with $1 < q \le 2$. Then $f \in W_0(\mathbb{R})$.

The simplest description of our main result of this section is that we extend the range of p and q in Theorem A as follows. This is not just a better machinery, any condition with $p > 2$ or $q > 2$ differs in principle and requires another approach.

Theorem 8.30 *Suppose a function f satisfies condition* **(N)**.

(a) *Let $f \in L^p(\mathbb{R})$, for some p, $1 \le p < \infty$, and $f'(t) \in L^q(\mathbb{R})$, for some q, $1 < q < \infty$. If*

$$\frac{1}{p} + \frac{1}{q} > 1,$$

then $f \in W_0(\mathbb{R})$.

(b) *If $\frac{1}{p} + \frac{1}{q} < 1$, then there exists a function f satisfying* **(N)** *such that $f(t) \in L^p(\mathbb{R})$ and $f'(t) \in L^q(\mathbb{R})$ but $f \notin W_0(\mathbb{R})$.*

As a corollary, we present a different proof of **I**).

Corollary 8.31 *If $\frac{\beta}{\alpha} > \frac{1}{2}$, then $m \in W_0(\mathbb{R})$.*

Using (a) in Theorem 8.30 and the Gagliardo–Nirenberg inequality, we can extend (a) to higher derivatives as follows.

Theorem 8.32 *Let $f \in C_0(\mathbb{R})$ and let f and its derivatives $f^{(j)}$, for all $j = 1, \ldots, r - 1$, be locally absolutely continuous on $\mathbb{R} \setminus \{0\}$. Let $f \in L^p(\mathbb{R})$, for some p, $1 \le p < \infty$, and $f^{(r)} \in L^s(\mathbb{R})$, for some r and s, $r > 1$, $1 < s < \infty$. If*

$$\frac{2r - 1}{p} + \frac{1}{s} > r, \tag{8.33}$$

then $f \in W_0(\mathbb{R})$.

We give, step by step, proofs of these assertions. We need certain auxiliary results. One of the basic tools is the following lemma (see Lemma 4 in [195] or Theorem 3 in [72], or maybe better [55, Theorem 6.4.2]; of course, in any dimension).

Lemma B *Let $f \in C_0(\mathbb{R})$. If*

$$\sum_{\nu=-\infty}^{\infty} 2^{\nu/2} \left(\int_{\mathbb{R}} |f(t + h(\nu)) - f(t - h(\nu))|^2 \, dt \right)^{\frac{1}{2}} < \infty,$$

where $h(\nu) = \pi 2^{-\nu}$, $\nu \in \mathbb{Z}$, then $f \in W_0(\mathbb{R})$.

This lemma is certainly a Bernstein type result, that is, a natural extension of the celebrated Bernstein's test for the absolute convergence of Fourier series (see [25, Chapter II, §6]) or Sect. 3.3 in Chap. 3.

A partial case of the Steklov–Hardy type inequality mentioned in the beginning of the previous chapter is as follows. For $F \ge 0$ and $1 < q \le Q < \infty$,

$$\left(\int_{\mathbb{R}} \left[\int_{t-h}^{t+h} F(s) \, ds \right]^Q dt \right)^{\frac{1}{Q}} \le Ch^{\frac{1}{Q} + \frac{1}{q'}} \left(\int_{\mathbb{R}} F^q(t) \, dt \right)^{\frac{1}{q}}. \tag{8.34}$$

As usual, $\frac{1}{q} + \frac{1}{q'} = 1$. Similarly $\frac{1}{p} + \frac{1}{p'} = 1$.

We will also apply the following simple result.

Lemma 8.35 *Let f satisfy* (**N**) *and $f' \in L^q(\mathbb{R})$. Then*

$$\|\Delta_h f\|_\infty \le 2^{\frac{1}{q'}} h^{\frac{1}{q'}} \|f'\|_q.$$

Proof By Hölder's inequality,

$$\|\Delta_h f\|_\infty \leq \int_{t-h}^{t+h} |f'(s)|\, ds \leq \left(\int_{t-h}^{t+h} ds \right)^{\frac{1}{q'}} \|f'\|_q,$$

as required. $\square$

8.3.1 Proof of the Main Theorem

The proof will be divided into several steps. Let us begin with simpler cases.

Step 1

To prove **(b)** of the theorem, let us consider the function m defined by (8.29). Suppose that $p\beta > 1$ and $q(\beta - \alpha + 1) > 1$, with $\alpha \neq 1$. Simple calculations show that $m \in L^p(\mathbb{R})$ and $m' \in L^q(\mathbb{R})$. If $\frac{\beta}{\alpha} < \frac{1}{2}$, then $m \notin A(\mathbb{R})$. The last inequality is equivalent to $2\beta - \alpha + 1 < 1$. Therefore,

$$\frac{1}{p} + \frac{1}{q} < 2\beta - \alpha + 1 < 1,$$

and the considered m delivers the required counterexample.

Step 2

Let $p, q \leq 2$, the known case from Theorem A. In order to show the difference in the methods, let us give a separate simple proof. In this case the Fourier transforms of both the function and its derivative can be understood in the sense of mean convergence (point-wise for f when $p = 1$), and hence there is a subsequence convergent almost everywhere. We will prove that $\widehat{f}$ is integrable. Indeed, let us estimate

$$\int_{\mathbb{R}} |\widehat{f}(x)|\, dx.$$

We split the integral into two, with $|x| \leq 1$ and otherwise. Applying Hölder's inequality and then the Hausdorff-Young inequality, we get

$$\int_{|x|\leq 1} |\widehat{f}(x)|\, dx \leq 2^{\frac{1}{p}} \left(\int_{\mathbb{R}} |\widehat{f}(x)|^{p'}\, dx \right)^{\frac{1}{p'}} \leq C_p \|f\|_p.$$

Further, let $|x| > 1$. Integrating by parts, we derive that for almost every x

$$|\widehat{f}(x)| \leq \frac{|\widehat{f'}(x)|}{|x|},$$

where the Fourier transform on the right exists whenever $f' \in L_q$, $q \leq 2$.

As above, applying Hölder's inequality and then the Hausdorff-Young inequality, we obtain

$$\int_{|x|>1} |\widehat{f}(x)| \, dx \leq \int_{|x|>1} \frac{|\widehat{f'}(x)|}{|x|} \, dx$$

$$\leq \left(\frac{2}{q-1} \right)^{\frac{1}{q}} \left(\int_{\mathbb{R}} |\widehat{f'}(x)|^{q'} \, dx \right)^{\frac{1}{q'}} \leq C_q \|f'\|_q,$$

which completes Step 2.

For more difficult cases of **(a)**, we apply Lemma B. Contrary to *Step 2*, we cannot make sure of the existence of the Fourier transform of f nor of f'. Correspondingly, we are not able to make use of the Hausdorff-Young inequality.

Denoting

$$\Delta(h) = \left(\int_{\mathbb{R}} |f(t+h) - f(t-h)|^2 dt \right)^{\frac{1}{2}}, \tag{8.36}$$

we are going to prove that

$$\sum_{v=1}^{\infty} 2^{-\frac{v}{2}} \Delta(h(-v)) + \sum_{v=0}^{\infty} 2^{\frac{v}{2}} \Delta(h(v)) < \infty. \tag{8.37}$$

It is obvious that for $h > 0$,

$$|f(t+h) - f(t-h)| = \left| \int_{t-h}^{t+h} f'(s) \, ds \right|. \tag{8.38}$$

Step 3

Let us start with the first sum in (8.37) which is

$$\sum_{v=1}^{\infty} 2^{-v/2} \left(\int_{\mathbb{R}} |f(t+h(-v)) - f(t-h(-v))|^2 dt \right)^{\frac{1}{2}}. \tag{8.39}$$

Using (8.38), we represent the integral as

$$\left(\int_{\mathbb{R}} |f(t+h(-v)) - f(t-h(-v))| \left| \int_{t-h(-v)}^{t+h(-v)} f'(s) \, ds \right| dt \right)^{\frac{1}{2}}.$$

By Hölder's inequality, it is estimated via

$$\left(\int_{\mathbb{R}} |f(t + h(-v)) - f(t - h(-v))|^p dt \right)^{\frac{1}{2p}}$$

$$\times \left(\int_{\mathbb{R}} \left[\int_{t-h(-v)}^{t+h(-v)} |f'(s)| ds \right]^{p'} dt \right)^{\frac{1}{2p'}}. \tag{8.40}$$

Since $p' > q$, we use (8.34) with $F(s) = |f'(s)|$ and $Q = p'$. Therefore, the first sum in (8.37) is controlled by

$$\| f \|_p^{1/2} \| f' \|_q^{\frac{1}{2}} \sum_{v=1}^{\infty} 2^{-\frac{v}{2}(1 - \frac{1}{p'} - \frac{1}{q'})},$$

and is bounded since

$$1 - \frac{1}{p'} - \frac{1}{q'} = \frac{1}{p} + \frac{1}{q} - 1 > 0.$$

Step 4
In order to estimate the second sum in (8.37) equal to

$$\sum_{v=0}^{\infty} 2^{\frac{v}{2}} \left(\int_{\mathbb{R}} |f(t + h(v)) - f(t - h(v))|^2 dt \right)^{\frac{1}{2}}, \tag{8.41}$$

we observe that the integral can be treated as follows.

Applying Hölder's inequality with the exponents $q' > 1$ and q, we estimate (8.41) via

$$\left(\int_{\mathbb{R}} |\Delta_{h(v)} f(t)|^{q'} dt \right)^{\frac{1}{2q'}} \left(\int_{\mathbb{R}} \left[\int_{t-h(v)}^{t+h(v)} |f'(s)| ds \right]^q dt \right)^{\frac{1}{2q}}. \tag{8.42}$$

By (8.38) and Lemma 8.35, the first integral in (8.42) is controlled by

$$\left(\int_{\mathbb{R}} |\Delta_{h(v)} f(t)|^p |\Delta_{h(v)} f(t)|^{q'-p} dt \right)^{\frac{1}{2q'}} \leq C h(v)^{\frac{q'-p}{q'} \frac{1}{2q'}} \| f \|_p^{\frac{p}{2q'}} \| f' \|_q^{\frac{q'-p}{2q'}}.$$

To estimate the second one, we use (8.34) with $F(s) = |f'(s)|$ and $Q = q$. We thus estimate the second factor in (8.42) via

$$C[h(v)]^{\frac{1}{2}} \| f' \|_q^{\frac{1}{2}}.$$

Since $q' > p$, the series

$$\sum_{\nu=1}^{\infty} 2^{-\nu \frac{q'-p}{q'} \frac{1}{2q'}}$$

converges, which ensures the finiteness of (8.37).

8.3.2   Proof of the Corollary

Similarly to (b) of the above theorem, we have $m \in L^p(\mathbb{R})$ for $p\beta > 1$ and $m' \in L^q(\mathbb{R})$ for $q(\beta - \alpha + 1) > 1$. If $\frac{\beta}{\alpha} > \frac{1}{2}$, the last inequality is equivalent to $2\beta - \alpha + 1 > 1$. Therefore,

$$\frac{1}{p} + \frac{1}{q} > 2\beta - \alpha + 1 > 1,$$

and $m \in W_0(\mathbb{R})$.

8.3.3   Proof of the Extended Theorem

The Gagliardo–Nirenberg inequality (with certain contribution by Ladyzhenskaya), see, e.g., [174], states, in dimension one, that if

$$\frac{1}{q} = j + \left(\frac{1}{s} - r\right)\alpha + \frac{1-\alpha}{p}, \tag{8.43}$$

with $\frac{j}{r} \leq \alpha \leq 1$, then

$$\|f^{(j)}\|_q \leq C \|f\|_p^{1-\alpha} \|f^{(r)}\|_s^{\alpha}. \tag{8.44}$$

Taking $j = 1$ and $\alpha = \frac{1}{r}$, we control the first derivative. Representing s from the corresponding version of (8.43), we obtain

$$\frac{1}{s} = \frac{r}{q} - \frac{r}{p} + \frac{1}{p}.$$

Inserting this $\frac{1}{s}$ in (8.33) results in

$$\frac{2r-1}{p} + \frac{r}{q} - \frac{r}{p} + \frac{1}{p} > r,$$

which is equivalent to $\frac{1}{p} + \frac{1}{q} > 1$. The assertion $f \in W_0(\mathbb{R})$ follows now from (a) in Theorem 8.30.

In fact, these results can by no means be sharp, since they claim for β to be positive, which is not necessary. This gap can be covered by assuming weighted type conditions. This is done in [135] for arbitrary dimension, while direct multivariate generalizations are proven in [134]. All these are beyond our consideration.

8.4 Boas' Conjecture

Integrability of the Fourier transforms with power weights induces natural constraints on the exponents. *Pitt's inequality* (see, e.g., [71]) asserts that

$$\left[\int_{-\infty}^{\infty} |\widehat{h}(x)|^p |x|^{-p\gamma} dx \right]^{\frac{1}{p}} \le C \left[\int_{-\infty}^{\infty} |h(t)|^q |t|^{q(1+\gamma-\frac{1}{p}-\frac{1}{q})} dt \right]^{\frac{1}{q}}$$

if $1 < q \le p < \infty$ and

$$\max\{0, \frac{1}{p} + \frac{1}{q} - 1\} < \gamma < \frac{1}{p};$$

the result fails for γ outside this range. To widen the range, additional restrictions on h should be posed. For instance, vanishing of certain moments is assumed in [184]. For monotone functions, R.P. Boas [75] conjectured the following. Let for $x \ge 0$,

$$H(x) = \int_0^{\infty} h(t) e^{ixt} dt. \qquad (8.45)$$

Conjecture *If* $1 < p < \infty$ *and* h *is non-negative and monotone on* $(0, \infty)$, *then*

$$x^{-\gamma} H(x) \in L^p(0, \infty) \qquad \text{if and only if} \qquad t^{1+\gamma-\frac{2}{p}} h(t) \in L^p(0, \infty)$$

provided $-\frac{1}{p'} = -1 + \frac{1}{p} < \gamma < \frac{1}{p}$.

In [162], an enriched version of Boas' conjecture is proved. First, weighted integrability of a function and its Fourier transform in different Lebesgue spaces, that is, in L^p and L^q is considered. Next, the result is obtained for a wider class of h being *general monotone* functions. For $p = q$, it is exactly the assertion of the conjecture but for a wider class of general monotone functions.

The notion of *general monotonicity* for functions is introduced in [163] as an analog of general monotonicity brought in for sequences in [193]. We call a function *admissible* if it is locally of bounded variation on $(0, \infty)$ and vanishes at infinity. The following definition is given in [163] (cf. [94]).

Definition 8.46 We call an admissible function h general monotone, written GM, if for all $x \in (0, \infty)$, there holds

$$\int_x^{2x} |dh(t)| \le C \int_{\frac{x}{c}}^{cx} \frac{|h(u)|}{u} \, du, \tag{8.47}$$

for some $c > 1$.

Note that any monotone (or *quasi-monotone*, that is, those positive h on $(0, \infty)$ for which $h(t)t^{-\alpha}$ is non-increasing for some positive α) function belongs to the GM class. We need the following two properties of the GM functions (see [163, 194]):

$$|h(x)| \le C \int_{\frac{x}{c}}^{cx} \frac{|h(u)|}{u} \, du, \qquad x > 0 \tag{8.48}$$

and

$$\int_x^y |dh(t)| \le C \int_{\frac{x}{c}}^{cy} \frac{|h(u)|}{u} \, du, \qquad 0 < x < y. \tag{8.49}$$

We will assume h to be integrable near the origin. By this, h is locally integrable on $\mathbb{R}_+$ and H is well-defined in the distributional sense. Our main result reads as follows.

Theorem 8.50 *Let h be a non-negative general monotone function on $(0, \infty)$. Let $1 \le p, q < \infty$ and $-\frac{1}{p'} < \gamma < \frac{1}{p}$.*

(A) *If $q \le p$, then*

$$x^{1+\gamma-\frac{1}{p}-\frac{1}{q}} h(x) \in L^q(0, \infty) \qquad \text{implies} \qquad x^{-\gamma} H(x) \in L^p(0, \infty);$$

(B) *If $p \le q$, then*

$$x^{-\gamma} H(x) \in L^p(0, \infty) \qquad \text{implies} \qquad x^{1+\gamma-\frac{1}{p}-\frac{1}{q}} h(x) \in L^q(0, \infty).$$

Proof We begin with the upper estimate of $\|x^{-\gamma} H(x)\|_{L^p}$. Let

$$\Phi(x) := \int_x^{2x} |dh(t)|.$$

Lemma 8.51 *There holds*

$$|H(x)| \le C \left(\int_0^{\frac{\pi}{x}} \Phi(t) \, dt + \frac{1}{x} \int_{\frac{\pi}{x}}^{\infty} \frac{\Phi(t)}{t} \, dt \right). \tag{8.52}$$

Proof Splitting the integral and integrating by parts, we obtain

$$
|H(x)| = \left| \int_0^{\frac{\pi}{x}} h(t)e^{ixt} dt + \frac{1}{ix} h\left(\frac{\pi}{x}\right) - \frac{ix}{\int} \int_{\frac{\pi}{x}}^{\infty} e^{ixt} dh(t) \right|
$$

$$
\leq \int_0^{\frac{\pi}{x}} |h(t)| + \frac{2}{x} \int_{\frac{\pi}{x}}^{\infty} |dh(t)|.
$$

Since

$$
\int_0^{\frac{\pi}{x}} |h(t)| \, dt \leq \int_0^{\frac{\pi}{x}} \int_t^{\frac{\pi}{x}} |dh(s)| \, dt + \int_0^{\frac{\pi}{x}} \int_{\frac{\pi}{x}}^{\infty} |dh(s)| \, dt,
$$

we have

$$
|H(x)| \leq C \int_0^{\frac{\pi}{x}} s \, |dh(s)| + \int_{\frac{\pi}{x}}^{\infty} |dh(s)|. \tag{8.53}
$$

Applying now, with ψ being any integrable function,

$$
\int_0^B \frac{1}{t} \int_t^{2t} |\psi(u)| \, du \, dt \geq \ln 2 \int_0^B |\psi(u)| \, du
$$

and

$$
\int_A^{\infty} \frac{1}{t} \int_t^{2t} |\psi(u)| \, du \, dt \geq \ln 2 \int_{2A}^{\infty} |\psi(u)| \, du
$$

to the right-hand side of (8.53), we complete the proof of the lemma.   □

We will use the following extension [81] of Hardy's inequality (it can be instructive to compare this version with those from the previous chapter): if $1 \leq \alpha \leq \beta < \infty$, then

$$
\left[\int_0^{\infty} u(x) \left(\int_0^x \psi(t) \, dt \right)^{\beta} dx \right]^{\frac{1}{\beta}} \leq C \left[\int_0^{\infty} v(x) \psi(x)^{\alpha} \, dx \right]^{\frac{1}{\alpha}} \tag{8.54}
$$

holds for every $\psi, u, v \geq 0$ if and only if

$$
\sup_{t>0} \left(\int_t^{\infty} u(x) \, dx \right)^{\frac{1}{\beta}} \left(\int_0^t v(x)^{1-\alpha'} \, dx \right)^{\frac{1}{\alpha'}} < \infty.
$$

Let $p \geq q$. Estimating the first integral on the right in (8.52), we wish to have

$$\left[\int_0^\infty x^{p\gamma-2}\left|\int_0^x \Phi(t)\,dt\right|^p dx\right]^{\frac{1}{p}} \leq C\left[\int_0^\infty x^{q\gamma+q-1-\frac{q}{p}}\Phi(x)^q\,dx\right]^{\frac{1}{q}},$$

which is true if and only if $\gamma < \frac{1}{p}$.

Estimating now the second integral on the right in (8.52), we obtain, by substituting $t \to \frac{\pi}{t}$ and applying (8.54) with appropriate power weights,

$$\left[\int_0^\infty x^{-p\gamma-p}\left|\int_{\frac{\pi}{x}}^\infty \frac{\Phi(t)}{t}\,dt\right|^p dx\right]^{\frac{1}{p}}$$

$$\leq \left[\int_0^\infty x^{-p\gamma-p}\left|\int_0^x \frac{1}{t}\Phi\left(\frac{\pi}{t}\right)dt\right|^p dx\right]^{\frac{1}{p}}$$

$$\leq C\left[\int_0^\infty x^{-q\gamma-1+\frac{q}{p}}\left(\frac{1}{x}\Phi\left(\frac{\pi}{x}\right)\right)^q dx\right]^{\frac{1}{q}}.$$

This holds only when $\gamma > -\frac{1}{p'}$.

Since h is GM, using (8.49) along with Hardy's inequality (8.54) with $\alpha = \beta = q$ yields

$$\left[\int_0^\infty |x^{-\gamma}H(x)|^p dx\right]^{\frac{1}{p}} \leq C\left[\int_0^\infty u^{q\gamma+q-1-\frac{q}{p}}\left(\int_{\frac{u}{c}}^{cu} \frac{h(s)}{s}\,ds\right)^q du\right]^{\frac{1}{q}}$$

$$\leq C\left[\int_0^\infty u^{q\gamma-1-\frac{q}{p}}\left(\int_0^u h(s)\,ds\right)^q du\right]^{\frac{1}{q}}$$

$$\leq C\left[\int_0^\infty u^{q\gamma+q-1-\frac{q}{p}}h(u)^q\,du\right]^{\frac{1}{q}},$$

with $\gamma < \frac{1}{p}$. This proves the first part of the theorem.

To prove part (B), we need appropriate lower estimates for $q \geq p$. Letting, in order to fit (8.45),

$$H_s(x) = \int_0^\infty h(t)\sin xt\,dt$$

to be the sine Fourier transform of h (with notation different from that in the previous matter), we have

$$\int_0^u H_s(x)\,dx = 2\int_0^\infty \frac{h(t)}{t}\sin^2\frac{ut}{2}\,dt \geq 0.$$

We then obtain, with c from (8.47),

$$\int_0^{\frac{\pi}{2cx}} |H(x)|\, dx \geq \int_0^{\frac{\pi}{2cx}} \left| H_s(x) \right| dx \geq \left| \int_0^{\frac{\pi}{2cx}} H_s(x)\, dx \right| \geq C \int_{\frac{x}{c}}^{cx} \frac{h(t)}{t}\, dt.$$

By this,

$$\left[\int_0^\infty \left(\int_{\frac{t}{c}}^{ct} \frac{h(s)}{s}\, ds \right)^q t^{q\gamma + q - 1 - \frac{q}{p}}\, dt \right]^{\frac{1}{q}}$$

$$\leq C \left[\int_0^\infty t^{q\gamma + q - 1 - \frac{q}{p}} \left(\int_0^{\frac{\pi}{2ct}} |H(u)|\, du \right)^q dt \right]^{\frac{1}{q}}.$$

Substituting $\frac{\pi}{2ct} = x$ and applying (8.54), we get

$$\left[\int_0^\infty x^{-q\gamma - q - 1 + \frac{q}{p}} \left(\int_0^x |H(u)|\, du \right)^q dx \right]^{\frac{1}{q}} \leq C \left(\int_0^\infty x^{-p\gamma} |H(x)|^p dx \right)^{\frac{1}{p}},$$

which is true if and only if $\gamma > -\frac{1}{p'}$. Applying (8.48), we have

$$\left[\int_0^\infty \left| h(t) t^{1 + \gamma - \frac{1}{p} - \frac{1}{q}} \right|^q dt \right]^{\frac{1}{q}} \leq C \left[\int_0^\infty \left| H(x) x^{-\gamma} \right|^p dx \right]^{\frac{1}{p}}, \qquad (8.55)$$

which completes the proof.   □

We note that part (B) of Theorem 8.50 holds if only $-\frac{1}{p'} < \gamma$.

In the case $p = q$, we have a direct generalization of Boas' conjecture.

Corollary 8.56 Let $1 \leq p < \infty$. If h is non-negative and general monotone on $(0, \infty)$, then

$$x^{-\gamma} H(x) \in L^p(0, \infty) \qquad \text{if and only if} \qquad t^{1 + \gamma - \frac{2}{p}} h(t) \in L^p(0, \infty)$$

provided $-\frac{1}{p'} < \gamma < \frac{1}{p}$.

We note that for monotone h this result was proved by Sagher [185]. In addition, our estimates of $\|x^{-\gamma} H(x)\|_{L^p}$ are similar to known estimates for sequences by Askey and Wainger [63].

We have considered the Fourier transform H to directly fit Boas' conjecture, since the latter does not distinguish between the cosine and sine transforms. It turns out that the difference between them is hidden in this way. Denote, in addition, the cosine Fourier transform of h by

$$H_c(x) = \int_0^\infty h(t) \cos xt \, dt,$$

which differs from the notation in the previous chapters in the same way and for the same reasons as above. It follows from Corollary 8.56 that for non-negative $h \in GM$, we have

$$x^{-\gamma} H_c(x) \in L^p(0, \infty) \qquad \text{if and only if} \qquad t^{1+\gamma-\frac{2}{p}} h(t) \in L^p(0, \infty)$$

for $-\frac{1}{p'} < \gamma < \frac{1}{p}$. However, the counterpart for H_s holds for the wider range $-\frac{1}{p'} < \gamma < 1 + \frac{1}{p}$.

Indeed, estimate (8.55), actually fulfilled for H_s, provides the part "only if" for $-\frac{1}{p'} < \gamma$. To improve $\gamma < \frac{1}{p}$, we consider an appropriate version of Lemma 8.51 for the sine transform alone:

$$|H_s(x)| \le C \left(x \int_0^{\frac{\pi}{x}} t\Phi(t) \, dt + \frac{1}{x} \int_{\frac{\pi}{x}}^\infty \frac{\Phi(t)}{t} \, dt \right),$$

which could be shown similarly to the proof of the lemma by using $|\sin(xt)| \le |xt|$. In addition, here we may assume a weaker condition of integrability of $th(t)$ near the origin. Further, by Hardy's inequality, there holds

$$\left[\int_0^\infty \left| x^{1-\gamma} \int_0^{\frac{\pi}{x}} t\Phi(t) \, dt \right|^p dx \right]^{\frac{1}{p}} \le C \left[\int_0^\infty u^{\gamma p + p - 2} \Phi^p(u) \, du \right]^{\frac{1}{p}}, \quad (8.57)$$

exactly with $\gamma < 1 + \frac{1}{p}$. The rest of the calculations are the same.

We finally mention here that $|x|^{-\gamma p}$ belongs to the Muckenhoupt A_p class for $-\frac{1}{p'} < \gamma < \frac{1}{p}$. Therefore, in the case of sine transform we can deal not only with Muckenhoupt weights. For definitions and the theory of these classes, see, e.g., [16, Chapter 9].

8.5   Salem Type Conditions

In Chapter II, §9 of [4], one can find a nice discussion on Salem type necessary conditions (appeared initially in [46]) for the trigonometric series (3.1)

$$\frac{a_0}{2} + \sum_{n=1}^\infty (a_n \cos nx + b_n \sin nx)$$

to be the Fourier series of an integrable function. More precisely, necessary conditions for a_n and b_n to be the Fourier coefficients are

$$\lim_{k\to\infty} k \sum_{n=1}^{\infty} \frac{a_n}{(k+\frac{1}{2})^2 - n^2} = 0 \tag{8.58}$$

and

$$\lim_{k\to\infty} \sum_{n=1}^{\infty} \frac{nb_n}{(k+\frac{1}{2})^2 - n^2} = 0, \tag{8.59}$$

respectively. A present day glance at these relations allows one to immediately recognize in the expressions under the limit sign in (8.58) and (8.59) the discrete even and odd Hilbert transforms, respectively. They differ slightly from (5.34) and (5.35) but in most problems are equivalent. Of course, the use of discrete Hilbert transforms to such extent was not the case at the time of Salem's or Bary's publication.

With such a possibility of relation between the Fourier expansion and Hilbert transform, we obtain necessary conditions for a function to belong to Wiener's algebra (see (4.7)).

8.5.1   Non-periodic Salem Conditions

Along with fact proven in [167] that the Hilbert transform of a function from Wiener's algebra exists at every point, necessary conditions can be given as (see [154])

Theorem 8.60 *If $f \in W_0(\mathbb{R})$, then its Hilbert transform $\mathcal{H}f(x)$ exists at every $x \in \mathbb{R}$ and $\lim_{|x|\to\infty} \mathcal{H}f(x) = 0$.*

Proof Expressing f in the Hilbert transform formula by means of (4.7), we wish to justify the change of the order of integration (cf. Sect. 5.5 in Chap. 5). For each $\delta > 0$, we have

$$\int_{|x-t|>\delta} \frac{1}{x-t} \int_{\mathbb{R}} g(u)e^{itu} du\, dt = \int_{\mathbb{R}} g(u) \int_{|x-t|>\delta} e^{itu} \frac{1}{x-t} dt\, du.$$

In order to insert the limit, as $\delta \downarrow 0$, under the sign of integration, we apply the Lebesgue dominated convergence test. For this, we prove that the integrand in u is dominated by a single integrable function independent of δ. Since g is integrable, it suffices to show that the integrals

$$\int_{|x-t|>\delta} e^{itu} \frac{1}{x-t} \, dt$$

are uniformly bounded. This is the case while integrating over $|x - t| \geq 1$. Substituting $x - t = s$, we see that the rest is equal to

$$\int_{\delta<|x-t|<1} \frac{e^{itu}}{x-t} \, dt = -ie^{ixu} \int_{\delta<|s|<1} \frac{\sin su}{s} \, ds. \tag{8.61}$$

It is well known that the integral on the right is uniformly bounded (cf. (5.30) in Chap. 5). This proves the existence everywhere of the Hilbert transform. Applying [27, Volume 2, Table 1.3 (3.1)], we get

$$\mathcal{H}e^{iu\cdot}(x) = -i \operatorname{signu} e^{iux}, \tag{8.62}$$

and the corresponding Fourier integral tends to zero by the Riemann–Lebesgue lemma, which completes the proof.   □

It is worth mentioning that using the criterion given before Theorem 6.55 in Chap. 6, relations (8.58) and (8.59) are derived from Theorem 8.60 in [168].

8.5.2 Applications

Let us derive from Theorem 8.60 a non-periodic analog of the result from [4, Chapter II, §9]. The latter says that for the series $\sum a_n \cos nt$ with $\{a_n\}$ tending to zero monotonously to be a Fourier series, it is necessary that

$$\lim_{n\to\infty} (a_n - a_{n+1}) \ln n = 0.$$

Theorem 8.63 *If the cosine Fourier transform of a monotone decreasing bounded function f is integrable, then*

$$\lim_{x\to\infty} \lim_{\delta\to 0+} (f(x-\delta) - f(x+\delta)) \ln \frac{x}{\delta} = 0. \tag{8.64}$$

Proof By (5.34), for the even case, the necessary condition in Theorem 8.60 easily reduces to

$$\lim_{x\to\infty} x \int_0^\infty \frac{f(t)}{x^2 - t^2} \, dt = 0.$$

We first prove that

$$\lim_{x \to \infty} x \int_{2x}^{\infty} \frac{f(t)}{x^2 - t^2} \, dt = 0.$$

Indeed, by monotonicity,

$$x \int_{2x}^{\infty} \frac{f(t)}{t^2 - x^2} \, dt \le x f(2x) x \int_{2x}^{\infty} \frac{dt}{t^2 - x^2}$$

$$\le \frac{4}{3} x f(2x) x \int_{2x}^{\infty} \frac{dt}{t^2} = \frac{2}{3} f(2x),$$

which tends to zero as $x \to \infty$. Now, the necessary condition that we restart with is

$$\lim_{x \to \infty} x \int_0^{2x} \frac{f(t)}{x^2 - t^2} \, dt = 0.$$

Further,

$$x \int_0^{2x} \frac{f(t)}{x^2 - t^2} \, dt = \frac{1}{2} \int_0^{2x} f(t) \left[\frac{1}{x - t} + \frac{1}{x + t} \right] dt.$$

Since

$$\int_0^{2x} \frac{f(t)}{x + t} \, dt \le \frac{1}{x} \int_0^{2x} f(t) \, dt$$

$$= \frac{1}{\sqrt{x}} \frac{1}{\sqrt{x}} \int_0^{\sqrt{x}} f(t) \, dt + \le \frac{1}{x} \int_{\sqrt{x}}^{2x} f(t) \, dt$$

$$\le \frac{1}{\sqrt{x}} f(0) + 2 f(\sqrt{x}),$$

where the right-hand side tends to zero as $x \to \infty$, we have to derive our result from

$$\lim_{x \to \infty} \text{P.V.} \int_0^{2x} \frac{f(t)}{x - t} \, dt = \lim_{x \to \infty} \lim_{\delta \to 0+} \int_{\delta}^x \frac{f(x - t) - f(x + t)}{t} \, dt = 0.$$

But, by monotonicity, the right-hand side is not smaller than

$$\lim_{x \to \infty} \lim_{\delta \to 0+} [f(x - \delta) - f(x + \delta)] \int_{\delta}^x \frac{dt}{t}, \tag{8.65}$$

which is equivalent to the left-hand side of the desired condition.   □

Remark 8.66 Conditions like (8.64) are mostly needed for constructing counterexamples. Sometimes weaker conditions accomplish the job. For example, instead of
passing to (8.65), one can similarly estimate from below by means of

$$
\lim_{x \to \infty} [f(x - a) - f(x + a)] \int_a^x \frac{dt}{t}
$$
$$
= \lim_{x \to \infty} [f(x - 1) - f(x + 1)] \ln x = 0, \tag{8.67}
$$

with $a > 0$ fixed.

This is enough, for example, to conclude that there are monotone functions with
non-integrable cosine Fourier transform. Taking such a function to be equal $\frac{1}{t}$ for

$$
2^{t^2} < x < 2^{(t+1)^2},
$$

we have

$$
[f(2^{t^2} - 1) - f(2^{t^2} + 1)] \ln 2^{t^2} = \left[\frac{1}{t - 1} - \frac{1}{t} \right] t^2 \ln 2 > \ln 2,
$$

and the limit as $t \to \infty$, which is equivalent to $x \to \infty$, can never be zero. This
construction is much less sophisticated than that in [54, 6.11, Theorem 125].

8.6 L^1 Convergence of Fourier Transforms

In Sect. 3.6 of Chap. 3 and partially in the previous section of this chapter, we have
touched on the issue of trigonometric series. More precisely, we have concentrated
on the problem of identifying whether a trigonometric series is the Fourier series of
an integrable function. However, there are two parallel classical topics in harmonic
analysis that started simultaneously more than 100 years ago in a work of Young
[200]. One of them is the mentioned problem whether a trigonometric series is a
Fourier series. Further, when we are within this scope and one of the corresponding
assumptions on the coefficients is valid, the second problem naturally arises. It reads
as follows:

Is it possible to approximate in L^1 *the function that the given series represents,
or just the series, by its partial sums?*

More precisely, what are the conditions on the coefficients for such a convergence? Even more precisely, to start studying the problem in question the series
already must be a Fourier series, otherwise there is nothing to be considered.
Therefore, the question mostly studied is what are *additional* assumptions on the
coefficients of the series, additional to those that guarantee the belonging to $\widehat{L^1}$, for

the L^1 convergence of the partial sums. Let us give an example of such a statement. Recall that the N-th partial sums of (3.35) are

$$s_N(f; x)_c = \frac{a_0}{2} + \sum_{k=1}^{N} a_k \cos kx. \tag{8.68}$$

A typical result can be outlined as follows.

Let the sequence of the coefficients of (3.35) be a bv null sequence. If it satisfies certain of the above mentioned assumptions so that (3.35) is the Fourier series of f, then

$$\lim_{N \to \infty} \int_0^\pi |s_N(f; x)_c - f(x)| \, dx = 0 \tag{8.69}$$

if and only if

$$\lim_{k \to \infty} a_k \ln k = 0. \tag{8.70}$$

Later on we shall give a more precise theorem to be generalized, for (3.36) as well, of the same form but with the assumptions specified. It is clear that, in fact, we can forget about the underlying problem for trigonometric series and start discussing the L^1 convergence issues for Fourier series. However, the two problems are so strongly tied that they do not allow one to forget about the first.

As it happens very often, our main goal is to find a non-periodic counterpart, that is, to extend the topic of L^1 convergence to the non-periodic case. Surprisingly, there is only one known attempt of such an extension, [111]. However, indeed this work cannot be considered as being within the scope. In [111], the partial integrals of the Fourier transform or their Cesàro means are analyzed. The problem that the authors of that work try to solve is, so to say, a problem of dual type, which needs different tools and leads to the results of a different form and nature. A direct generalization should be read as follows.

Like for the series, we have already seen that the cosine Fourier transform (4.5) and the sine Fourier transform (4.6) and their integrability properties are more convenient to be studied rather than those of the general Fourier transform. We shall deal with functions f of bounded variation on $\mathbb{R}_+ = [0, \infty)$, and vanishing at infinity, $\lim_{t \to \infty} f(t) = 0$, written $f \in BV_0[0, \infty)$, and locally absolutely continuous on $(0, \infty)$, written $f \in LAC(0, \infty)$. As a whole, we will denote these by

$$f \in BV_0[0, \infty) \cap LAC(0, \infty).$$

By this, the right-hand sides of (4.5) and (4.6), with such f, are natural analogs of trigonometric series (3.35) and (3.36), respectively, with bv coefficients. The recent book [34] gives more or less a complete picture of assumptions on $f \in BV_0[0, \infty) \cap$

$LAC(0, \infty)$ under which the Fourier transforms in (4.5) and (4.6) are Lebesgue integrable. Along with the fact that the right-hand sides of (4.5) and (4.6) converge almost everywhere to their left-hand sides, one may start studying the problems of the L^1 convergence of the partial integrals

$$S_N(f; x)_c = \int_0^N f(t) \cos xt \, dt \tag{8.71}$$

and

$$S_N(f; x)_s = \int_0^N f(t) \sin xt \, dt, \tag{8.72}$$

to $\widehat{f_c}(x)$ and $\widehat{f_s}(x)$, respectively.

As is explained above, we first have to be sure that $\widehat{f_c}$ or $\widehat{f_s}$ are Lebesgue integrable. Various interesting sufficient conditions for this are collected in [34]. However, there is a single necessary and sufficient condition for this. Following [126], we introduce the space

$$Q = \{g \in L^1(\mathbb{R}) : \int_{\mathbb{R}} \frac{|\widehat{g}(x)|}{|x|} \, dx < \infty\},$$

a subspace of $L^1(\mathbb{R})$, or, more precisely, of $L_0^1(\mathbb{R})$—the space of $L^1(\mathbb{R})$ functions with mean zero. The integral in the definition of Q is dominated by the functions g in the real Hardy space (by the Fourier-Hardy inequality (5.27)). However, Q is a properly intermediate space between the real Hardy space and the $L_0^1(\mathbb{R})$. The mentioned criterion is given by

Theorem 8.73 *Let f be absolutely continuous on $\mathbb{R}$ and vanishing at infinity. Then $\widehat{f} \in L^1(\mathbb{R})$ if and only if $f' \in Q$.*

If reduced to the half-axis, we have the same results. However, continuing to keep the notation Q, we recall that, in fact, in each of the two cases Q is reduced to either its subspace of even functions or to the subspace of odd functions, with $\widehat{f_c}$ or $\widehat{f_s}$ on the half-axis in place of $\widehat{g}$ on $\mathbb{R}$. In fact, the result itself is just integration by parts, while numerous delicate conditions, sometimes sophisticated, can be found in [34]. However, Q is an interesting space, wider than all the other spaces of integrability of the Fourier transform of a function of bounded variation.

In most applications we have to level down the claims for integrability conditions, that is, to concentrate on certain sufficient conditions. Among the variety of such conditions, or subspaces of Q considered in Sect. 6.4 of Chap. 6, we choose O_q and E_q, $1 < q < \infty$. Denoting (like in the proof of Theorem 8.4 but on the half-axis rather than on the whole $\mathbb{R}$)

$$T_q g(x) := \left(\frac{1}{x} \int_x^{2x} |g(t)|^q dt \right)^{\frac{1}{q}},$$

with natural $T_\infty g(x) := \operatorname*{ess\,sup}_{x \le t \le 2x} |g(t)|$ for $q = \infty$, we can rewrite the norm in O_q, $1 < q \le \infty$, as

$$\|g\|_{O_q} = \int_0^\infty T_q g(x)\,dx.$$

Recall that

$$\|f\|_{E_q} = \|f'\|_{O_q} + \int_0^\infty \frac{|f(t)|}{t}\,dt.$$

Of course, if $f' \in O_q$, then E_q is a part of such functions. We will need the following result (see, e.g., [141]; there and in [34] as well as in some other sources, the sine part is given in a more precise, asymptotic form).

Theorem 8.74 *Let $f \in BV_0[0, \infty) \cap LAC(0, \infty)$ and $f' \in O_q(\mathbb{R}_+)$ for some $1 < q \le \infty$. Then the cosine Fourier transform of f is integrable, with*

$$\|\widehat{f_c}\|_{L^1(\mathbb{R}_+)} \lesssim \|f'\|_{O_q(\mathbb{R}_+)};$$

if $f \in E_q(\mathbb{R}_+)$ for some $1 < q \le \infty$, then the sine Fourier transform of f is integrable, with

$$\|\widehat{f_s}\|_{L^1(\mathbb{R}_+)} \lesssim \|f\|_{E_q(\mathbb{R}_+)}.$$

We now have all the needed prerequisites.

8.6.1 L^1 Convergence

We are now in a position to formulate our main result. Since the approximated Fourier transform is supposed to be integrable, there is no need to approximate it near infinity, where it is already L^1 small. Therefore, a natural analog (and extension) of the periodic case reads as follows.

Theorem 8.75 *Let $f \in BV_0[0, \infty) \cap LAC(0, \infty)$. If $f' \in O_q(\mathbb{R}_+)$ for some $1 < q \le \infty$, then, for any fixed $M > 0$,*

$$\lim_{N \to \infty} \int_0^M |S_N(f; x)_c - \widehat{f_c}(x)|\,dx = 0$$

if and only if

$$\lim_{t \to \infty} |f(t)| \ln t = 0. \tag{8.76}$$

If $f \in E_q(\mathbb{R}_+)$ for some $1 < q \leq \infty$, then, for any fixed $M > 0$,

$$\lim_{N \to \infty} \int_0^M |S_N(f;x)_s - \widehat{f_s}(x)| \, dx = 0$$

if and only if (8.76) holds true.

Proof A typical way to establish the L^1 convergence for trigonometric series is to start with the Fejér means. We implement this expedient for the Fourier transforms as well. Here the Fejér means are

$$\sigma_N(f;x)_c = \int_0^N \left(1 - \frac{t}{N}\right) f(t) \, \cos xt \, dt$$

and

$$\sigma_N(f;x)_s = \int_0^N \left(1 - \frac{t}{N}\right) f(t) \, \sin xt \, dt.$$

They are known to converge in the L^1 norm. By this, we have to search for conditions that ensure

$$\lim_{N \to \infty} \|S_N(f;\cdot)_c - \sigma_N(f;\cdot)_c\|_{L^1} = 0, \tag{8.77}$$

and similarly for $\sigma_N(f;\cdot)_s$. Thus, (8.77) can be rewritten and reformulated as under what conditions we have

$$\lim_{N \to \infty} \int_{\mathbb{R}_+} \left| \int_0^N f(t) \frac{t}{N} \cos tx \, dt \right| dx = 0, \tag{8.78}$$

or when an analogous relation (zero limit) for the sine case holds.

We first prove the cosine case. Taking into account the above arguments, we have to prove (8.78). Integration over $[0, \frac{1}{N}]$ reduces to

$$\int_0^{\frac{1}{N}} \left| \int_0^N f(t) \frac{t}{N} \cos tx \, dt \right| dx \leq \int_0^{\frac{1}{N}} \int_0^N |f(t)| \, dt \, dx$$

$$\leq \frac{1}{N} \int_0^{\sqrt{N}} |f(t)| \, dt + \max_{\sqrt{N} \leq t \leq N} |f(t)|,$$

which tends to zero as $N \to \infty$.

The estimate over $[\frac{1}{N}, M]$ is more delicate. Integrating by parts, we obtain

$$\int_0^N f(t)\frac{t}{N}\cos tx\, dt = f(t)\left[\frac{1}{x}\frac{t}{N}\sin tx + \frac{\cos tx - 1}{Nx^2}\right]\Big|_0^N$$

$$-\frac{1}{x}\int_0^N f'(t)\frac{t}{N}\sin tx\, dt - \int_0^N f'(t)\frac{\cos tx - 1}{Nx^2}\, dt.$$

The integrated terms are

$$f(N)\left[\frac{\sin Nx}{x} + \frac{\cos xN - 1}{Nx^2}\right] := J_1 + J_2,$$

while the rest can be rewritten as

$$-\frac{1}{x}\left[\int_0^N f'(t)\sin tx\, dt - \int_0^N f'(t)\left(1 - \frac{t}{N}\right)\sin tx\, dt\right]$$

$$-\int_0^N f'(t)\frac{\cos tx - 1}{Nx^2}\, dt := I_1 + I_2.$$

Estimating J_1 leads to integrating over $\frac{1}{N} \leq x \leq M$, which is equivalent to

$$|f(N)|\int_{\frac{1}{N}}^M \frac{dx}{x},$$

and letting this tend to zero is equivalent to (8.76). The same integration of J_2 is equivalent to $|f(N)|$, which tends to zero by assumption.

Let us proceed to estimating I_1 and I_2. In I_1, there are two integrals. However, by the known properties of the Fourier transform of an integrable function (recall that f' is integrable), both tend to the same limit $\widehat{f'_s}$ uniformly. Since $\dfrac{\widehat{f'_s}(x)}{x}$ is integrable ($f' \in Q$), we can change the passage to the limit as $N \to \infty$ and integration, which makes I_1 tend to zero. The estimate of I_2 obviously reduces to

$$\frac{1}{N}\int_{\frac{1}{N}}^M x^{-\frac{3}{2}}\int_0^N |f'(t)|\sqrt{t}\, dt\, dx.$$

Since for a nonnegative function ϕ we have

$$\int_0^N \frac{1}{t}\int_t^{2t} \phi(u)\, du\, dt = \ln 2\int_0^N \phi(u)\, du + \int_N^{2N} \phi(u)\ln\frac{2N}{u}\, du,$$

there holds

$$\int_0^N |f'(t)|\sqrt{t}\, dt \lesssim \int_0^N \frac{1}{\sqrt{t}} \int_t^{2t} |f'(u)|\, du\, dt.$$

Applying Hölder's inequality to the inner integral on the right, we obtain the bound, for $q > 1$,

$$\int_0^N \sqrt{t} T_q f'(t)\, dt.$$

After integrating in x, we have to estimate

$$\frac{1}{\sqrt{N}} \int_0^N \sqrt{t} T_q f'(t)\, dt.$$

Splitting the integral in two, over $[0, \sqrt{N}]$ and over $[\sqrt{N}, N]$, we get that the first one is dominated by

$$N^{-\frac{1}{4}} \|f'\|_{O_q},$$

which tends to zero as $N \to \infty$. For the rest, we obtain

$$\int_{\sqrt{N}}^N T_q f'(t)\, dt$$

to be estimated. Since $T_q f'(t)$ is integrable, this integral tends to zero as $N \to \infty$, which completes the proof for the cosine transform. The proof for the sine transform is very much similar. The only real difference is that when starting the estimates for integration over $[\frac{1}{N}, M]$ for the cosine transform, we are sure of the value $f(0)$, since the integrated terms in the integration by parts vanished at $t = 0$ "automatically". Here we use that $f(0) = 0$, which follows from the second condition in the assumption $f \in E_q(\mathbb{R}_+)$. It remains to mention that in the estimates for $\sin tx$, we use exactly the same bounds as for $\cos tx - 1$. $\square$

8.6.2 Application to Trigonometric Series

Given a cosine series and a sine series with the null sequence of bv coefficients each. Among numerous conditions for these trigonometric series to be a Fourier series, we concentrate on the sequence prototype of Theorem 8.74, see **(2)** in Sect. 3.6. We will rewrite it in an appropriate form. For this, we need the prototypes of the O_q

and E_q spaces, written o_q and e_q, respectively. The first one will be defined, for $1 < q < \infty$, by

$$\|d\|_{o_q} = \sum_{n=0}^{\infty} 2^{\frac{n}{q'}} \left\{ \sum_{k=2^n}^{2^{n+1}-1} |\Delta d_k|^q \right\}^{\frac{1}{q}} < \infty, \qquad \frac{1}{q} + \frac{1}{q'} = 1, \tag{8.79}$$

and for $q = \infty$, by

$$\|d\|_{o_\infty} = \sum_{k=0}^{\infty} \sup_{n \le k < 2n} |d_n|. \tag{8.80}$$

A similar but somewhat different notation for this can be found in (3.41) and (3.42) in Chap. 3. Similarly to the above, for $1 < q \le \infty$,

$$\|d\|_{e_q} = \|\{\Delta d_k\}\|_{o_q} + \sum_{k=1}^{\infty} \frac{|d_k|}{k}.$$

Theorem 8.81 *Given cosine and sine series with the null sequence of bv coefficients. If the sequence $\{\Delta a_k\}$ belongs to o_q, then (3.35) is a Fourier series, and if $\{b_k\}$ belongs to e_q, then (3.36) is a Fourier series; in both cases we denote by f the function which the series represents and converges to.*

This result (in a different notation) was proved in [105] along with

Theorem 8.82 *Let the coefficients of (3.35) form a bv null sequence such that $\{\Delta a_k\}$ belongs to o_q for some $1 < q \le \infty$. Then (8.69) holds if and only if (8.70) is valid. If the coefficients of (3.36) form a bv null sequence such that $\{b_k\}$ belongs to e_q for some $1 < q \le \infty$, then*

$$\lim_{N \to \infty} \int_0^\pi |s_N(f; x)_s - f(x)| \, dx = 0 \tag{8.83}$$

if and only if (8.70) is valid.

To derive Theorem 8.81 from Theorem 8.74, or any similar theorem, a "bridge" for passing from Fourier integrals to trigonometric series (and vice versa) is used, see (6.50) in Sect. 6.7 and further explanations. In other words, we construct a corresponding zigzag function by means of linear interpolation of the sequence of coefficients. The rest is simple: such a function inherits all the properties of the sequence and fits the conditions of Theorem 8.74, then the assertions of the latter can be rewritten in terms of the sequence, again by (6.50), which implies Theorem 8.81.

It turns out that the same approach is applicable to the problems of L^1 convergence. Indeed, considering in the proof of (6.50) (see [55, 4.1.2] or [34, Chapter 8]) the sum over $N \le k < \infty$ and the integral over $[N, \infty)$, we get the same formula with the variation of f over $[N, \infty)$. Since for the functions considered it tends to zero as $N \to \infty$, we can see that the sum of the series is approximated by the partial sums if and only the Fourier transform is approximated by the partial integrals over $[0, \pi]$. In symbols,

$$\lim_{N \to \infty} \sup_{|x| \leq \pi} \left| \int_{|t| \geq N} f(t)e^{ixt} dt - \sum_{|k| \geq N} f(k)e^{ixk} \right| = 0. \tag{8.84}$$

We are now in a position to fulfil all the technicalities for a new proof of Theorem 8.82, that is, to derive it from Theorem 8.75.

Proof of Theorem 8.82 For the same reasons as above, we give explicitly only the proof of the cosine case. The derivative of the corresponding zigzag is a step-wise function with the values $\{\Delta a_k\}$ almost everywhere. Therefore, the belonging of this sequence to o_q implies the belonging of the function to O_q. For this function, (8.76), necessary and sufficient for the L^1 convergence of the series, is obviously equivalent to (8.70), necessary and sufficient for the L^1 convergence on $[0, \pi]$ of the partial Fourier integrals. Of course, the L^1 convergence of the series follows from (the corresponding version for cosines of) (8.84). $\square$

Of course, the results of this section (and [157]) has only gotten to first base in the new extension of the old topic. Hopefully, further steps will follow. Among the results to be extended to the non-periodic case, Theorem 4 in [64] and some other (see the discussion in Sections 3.2 and 3.3 of [107] by Fridli as well as his paper [108]).

8.7 More About Applications

This brief section plays the role of an epilogue for this chapter and, in a sense, for the whole book. However, it cannot eliminate possible questions that the reader would like to ask after reading or looking through the text. The last (but not least) of such questions might be why the treatment of facts and methods is terminated at this very moment. Many answers can be found but none of them will be wholly satisfactory. Of course, we definitely can present more problems and material for this chapter (as well as for the other ones). However, enough's enough, more is less. Several important topics have been covered in this last chapter in addition to the main corpus. On the other hand, the author has undertaken tremendous efforts to convince the reader (and himself) that there is no need for some other topics here, from both logical and aesthetical points of view. In defence of such a position, the author is convinced that there are interesting open problems that can be solved by using only the tools given in the preceding chapters. Of course, this is not as an end in itself. It will be a real pleasure to face an obstacle while solving some problem within that scope for which something beyond the considered machinery is needed, to solve such a problem and thus to expand the stock of his knowledge and his abilities (know-how). A path has been outlined in these chapters. To make the path longer or to discover a different one could be the best award for curiosity. As the saying goes "Live and learn". But this would be, should this be the case, material for another book or other papers.

Basic Notations

We present the list of basic notations in the order in which they appear in the text of the book. Some notations are not listed if they are used "locally", that is, in only a specific certain section or subsection.

$\lesssim$ and $\gtrsim$	Abbreviations for $\leq C$ and $\geq C$, where $C > 0$ is a constant
$\asymp$	Abbreviation for equivalence bilateral estimate
$\omega(f; h)$, $\omega(f; h)_p$	Modulus of continuity
$\omega_k(f; h)$	modulus of smoothness
Lip 1, **Lip** α	Lipschitz classes
f^*, d^*	Decreasing rearrangement
$\|f\|_{p,v}$	Norm in the weighted Lebesgue space
$Vf := V_{[a,b]}f$	Total variation of f on $[a, b]$
$f \in BV([a, b])$	f is of bounded variation on $[a, b]$
BV_0	Functions of bounded variation vanishing at infinity
$AC([a, b])$	Absolutely continuous functions on $[a, b]$
$LAC([a, b])$	Locally absolutely continuous functions on $[a, b]$
bv	Sequences of bounded variation
$Mf(x) = \sup\limits_{r>0} \frac{1}{2r} \int_{|t-x|<r} |f(t-x)| \, dt$	Hardy–Littlewood maximal function
$\widehat{g}(x) = \int_{-\infty}^{\infty} g(t)e^{-ixt} \, dt$	Fourier transform
$\check{h}(t) = \frac{1}{2\pi} \int_{-\infty}^{\infty} h(t)e^{ixt} \, dx$	Inverse Fourier transform
$\widehat{f}_c(x) = \int_0^{\infty} f(t)\cos xt \, dt$	Cosine Fourier transform
$\widehat{f}_s(x) = \int_0^{\infty} f(t)\sin xt \, dt$	Sine Fourier transform
$W_0(\mathbb{R}) = \{f \in C_0(\mathbb{R}) : f(t) = \int_{\mathbb{R}} g(x)e^{itx} \, dx, g \in L^1(\mathbb{R})\}$	Wiener algebra
$\|\{d_j\}\|_{\ell^p} = \left(\sum\limits_{j=-\infty}^{\infty} |d_j|^p \right)^{\frac{1}{p}}$	Norm in the space of p-summable sequences
$W(L^p, \ell^q)$	Wiener amalgam space

© The Author(s), under exclusive license to Springer Nature Switzerland AG 2021
E. Liflyand, *Harmonic Analysis on the Real Line*, Pathways in Mathematics,
https://doi.org/10.1007/978-3-030-81892-0

$$\|f\|_{(p,q)} = \left(\sum_{j=-\infty}^{\infty} \left[\int_j^{j+1} |f(t)|^p \, dt \right]^{\frac{q}{p}} \right)^{\frac{1}{q}} dx$$ Norm in the $W(L^p, \ell^q)$

$\mathcal{S}(\mathbb{R})$ Schwartz class

$\mathcal{S}'$ Space of tempered distributions

$\mathcal{H}g(x) = \frac{1}{\pi} \int_{\mathbb{R}} \frac{g(t)}{x-t} \, dt$ Hilbert transform

$\mathcal{H}_\delta g(x) = \frac{1}{\pi} \int_{|t-x|>\delta} \frac{g(t)}{x-t} \, dt$ Truncated Hilbert transform

$\mathcal{H}^* g(x) = \sup_{\delta>0} |\mathcal{H}_\delta g(x)|$ Maximal Hilbert transform

$\widetilde{\mathcal{H}} f(x) = \text{P.V.} \frac{1}{\pi} \int_{\mathbb{R}} f(t) \left\{ \frac{1}{x-t} + \frac{t}{1+t^2} \right\} dt$ Modified Hilbert transform

$\mathcal{H}_o g(x) = \frac{2}{\pi} \int_0^\infty \frac{t g(t)}{x^2-t^2} \, dt$ Odd Hilbert transform

$\mathcal{H}_e g(x) = \frac{2}{\pi} \int_0^\infty \frac{x g(t)}{x^2-t^2} \, dt$ Even Hilbert transform

$P_t(x) = \frac{1}{\pi} \frac{t}{t^2+x^2}$ Poisson kernel

$Q_t(x) = \frac{1}{\pi} \frac{x}{t^2+x^2}$ Conjugate Poisson kernel

$\hbar a(m) = \sum_{\substack{k=-\infty \\ k \neq m}}^{\infty} \frac{a_k}{m-k}$ Discrete Hilbert transform

$\hbar_o a(m) = \sum_{\substack{k=1 \\ k \neq m}}^{\infty} \frac{2k a_k}{m^2-k^2} - \frac{a_m}{2m}$ Odd discrete Hilbert transform

$\hbar_e a(m) = \sum_{\substack{k=1 \\ k \neq m}}^{\infty} \frac{2m a_k}{m^2-k^2} + \frac{a_m}{2m}$ Even discrete Hilbert transform

$L_0^1(\mathbb{R})$ Space of wavelet functions

$L \log L = \{f : \int_{\mathbb{R}} |f(t)| \ln^+ |f(t)| \, dt < \infty\}$ Zygmund class

$\|g\|_{H^1(\mathbb{R})} = \|g\|_{L^1(\mathbb{R})} + \|\mathcal{H}g\|_{L^1(\mathbb{R})}$ Norm of g in the real Hardy space

$H_o^1(\mathbb{R}_+)$ Subspace of odd functions in the real Hardy space

$H_e^1(\mathbb{R}_+)$ Subspace of even functions in the real Hardy space

$\|g\|_{O_q} = \int_0^\infty \left(\frac{1}{x} \int_x^{2x} |g(t)|^q \, dt \right)^{\frac{1}{q}} dx, 1 < q < \infty,$

$\|g\|_{O_\infty} = \int_0^\infty \underset{x \leq t \leq 2x}{\text{ess sup}} |g(t)| \, dx$ Norms in the subspaces of $H_o^1(\mathbb{R}_+)$

$L^* := L^*(\mathbb{R}) = \{f : \|f\|_{L^*} = \int_0^\infty \underset{|t| \geq x}{\text{ess sup}} |f(t)| \, dx < \infty\}$

$E_q := E_q(\mathbb{R}_+) := \left\{ g \in O_q : \int_0^\infty \frac{1}{x} \left| \int_0^x g(t) \, dt \right| \, dx < \infty \right\}$ Subspaces of $H_e^1(\mathbb{R}_+)$

$Q = \{g : g \in L^1(\mathbb{R}), \int_{\mathbb{R}} \frac{|\hat{g}(x)|}{|x|} \, dx < \infty\}$ Widest space for the integrability of the Fourier transform of a function of bounded variation

$$\|g\|_{A_{1,2}} = \sum_{m=-\infty}^{\infty} \left\{ \sum_{j=1}^{\infty} \left[\int_{j2^m}^{(j+1)2^m} |g(t)| \, dt \right]^2 \right\}^{\frac{1}{2}} dx < \infty$$ Norm of g in $A_{1,2}$

$\mathcal{K}f := \mathcal{K}_\varphi f$ Hausdorff operator

$$K(f, t; X_1, X_2) = \inf_{f=g+h}(\|g\|_{X_1} + t\|h\|_{X_2}) \qquad K\text{-functional}$$

$$\|d\|_{o_q} = \sum_{n=0}^{\infty} 2^{\frac{n}{q'}} \left\{ \sum_{k=2^n}^{2^{n+1}-1} |\Delta d_k|^q \right\}^{\frac{1}{q}} < \infty,$$

$$1 < q < \infty, \quad \frac{1}{q} + \frac{1}{q'} = 1$$

$$\|d\|_{o_\infty} = \sum_{k=0}^{\infty} \sup_{n \leq k < 2n} |d_n|$$

$$\|d\|_{e_q} = \|\{\Delta d_k\}\|_{o_q} + \sum_{k=1}^{\infty} \frac{|d_k|}{k}$$

$$\|d\|_{h^1} = \|g\|_{\ell^1} + \|\hbar d\|_{\ell^1} \qquad \text{Norm of } \{d\} \text{ in the discrete Hardy space}$$

h_o^1 — Subspace of odd sequences in the discrete Hardy space

h_e^1 — Subspace of even sequences in the discrete Hardy space

Bibliography

Textbooks and Monographs

1. N.I. Akhiezer, *The Classical Moment Problem and Some Related Questions in Analysis* (Oliver Boyd, 1965)
2. G. Alexits, *Convergence Problems of Orthogonal Series* (Akadémiai Kiadó, Budapest, 1961)
3. J. Appell, J. Banas, N.J. Merentes Diáz, *Bounded Variation and Around*. De Gruyter Series in Nonlinear Analysis and Applications (De Gruyter, 2013)
4. N.K. Bary, *A Treatise on Trigonometric Series, I and II* (MacMillan, New York, 1964)
5. H. Bateman, A. Erdélyi , *Tables of Integral Transforms*, vol. II (McGraw Hill Book Company, New York, 1954)
6. C. Bennett R. Sharpley, *Interpolation of Operators* (Academic Press, New York, 1988)
7. J. Bergh, J.Löfström, *Interpolation Spaces. An Introduction* (Springer, Berlin-New York, 1976)
8. S.V. Bochkarev, *A Method of Averaging in the Theory of Orthogonal Series and Some Problems in the Theory of Bases*. Trudy Inst. Steklova, vol. 146 (1978) (Russian). English transl. in Proc. Steklov Inst. Math. **3**, 1980
9. S. Bochner, *Lectures on Fourier Integrals* (Princeton Univ. Press, Princeton, 1959)
10. P.L. Butzer, R.J. Nessel, *Fourier Analysis and Approximation. Volume 1. One-Dimensional Theory* (Academic Press, New York and London, 1971)
11. E.M. Dyn'kin, in *Methods of the Theory of Singular Integrals: Hilbert Transform and Calderón-Zygmund Theory*, ed. by V.P. Khavin, N.K. Nikol'skij. Commutative Harmonic Analysis I (Springer, Berlin, 1991), pp. 167–259
12. J. Duoandikoetxea, *Fourier Analysis*. Graduate Studies in Mathematics, vol. 29 (Amer. Math. Soc., Providence, 2001)
13. J. Garcia-Cuerva, J.L. Rubio de Francia, *Weighted Norm Inequalities and Related Topics* (North-Holland, Amsterdam, 1985)
14. J.B. Garnett, *Bounded Analytic Functions* (Springer, New York, 2007)
15. R.R. Goldberg, *Fourier Transforms* (Cambridge Univ. Press, New York, 1961)
16. L. Grafakos, *Modern Fourier Analysis*, 2nd ed. (Springer, Berlin, 2009)
17. K. Gröchenig, *Foundations of Time-Frequency Analysis*. Appl. Numer. Harmon. Anal. (Birkhäuser, Boston, 2001)
18. N.B. Haaser, J.A. Sullivan, *Real Analysis*. Revised reprint of the 1971 original (Dover Publications, New York, 1991)

© The Author(s), under exclusive license to Springer Nature Switzerland AG 2021
E. Liflyand, *Harmonic Analysis on the Real Line*, Pathways in Mathematics,
https://doi.org/10.1007/978-3-030-81892-0

19. P.R. Halmos, *I Want to be a Mathematician. An Automathography in Three Parts*. MAA Spectrum (Mathematical Association of America, Washington, 1985)
20. E.W. Hansen, *Fourier Transforms: Principles and Applications* (Wiley, Hoboken, 2014)
21. G.H. Hardy, *Divergent Series* (Clarendon Press, Oxford, 1949)
22. G.H. Hardy, J.E. Littlewood, G. Pólya, *Inequalities*, Cambridge Math. Library, 2nd ed. (Cambridge Univ. Press, Cambridge, 1952)
23. E. Hewitt, K. Stromberg, *Real and Abstract Analysis* (Springer, Heidelberg/Berlin, 1965)
24. A. Iosevich, E. Liflyand, *Decay of the Fourier Transform: Analytic and Geometric Aspects* (Birkhäuser, Boston, 2014)
25. J.-P. Kahane, *Séries de Fourier Absolument Convergentes* (Springer, Berlin, 1970)
26. T. Kawata, *Fourier Analysis in Probability Theory* (Academic Press, New York, 1972)
27. F.W. King, *Hilbert Transforms*, vol. 1. Enc. Math/ Appl. (Cambridge Univ. Press, Cambridge, 2009)
28. A.W. Knopp, *Basic Real Analysis*. Digital Second Edition (Published by the Author, East Setauket, New York, 2016)
29. A.N. Kolmogorov, S.V. Fomin, in *Elements of the Theory of Functions and Functional Analysis*, ed. by V.M. Tikhomirov, 5th edn. With a supplement "Banach algebras" (Nauka, Moscow, 1981) (Russian)
30. P. Koosis, *Introduction to H_p Spaces* (Cambridge Univ. Press, Cambridge, 1980)
31. M.G. Krein, A.A. Nudelman, *The Markov Moment Problem and Extremal Problems. Ideas and Problems of P. L. Chebyshev and A. A. Markov and Their Further Development*. Translations of Mathematical Monographs, vol. 50 (American Mathematical Society, Providence, 1977)
32. A. Kufner, L. Maligranda, L.-E. Persson, *The Hardy Inequality. About its History and Some Related Results* (Vydavatelsk Servis, Plzen, 2007)
33. A. Kufner, L.-E. Persson, *Weighted Inequalities of Hardy Type* (World Scientific Publishing, River Edge, 2003)
34. E. Liflyand, *Functions of Bounded Variation and their Fourier Transforms* (Birkhäuser, Boston, 2019)
35. E. Lukacs, *Characteristic Functions*, 2nd edn. (Charles Griffin, London, 1970)
36. N.N. Luzin, *Integral and Trigonometric Series*. Editing and Commentary by N. K. Bari and D. E. Men'shov. Gosudarstv. Izdat. Tehn.-Teor. Lit., Moscow-Leningrad, 1951 (Russian)
37. B.M. Makarov, M.G. Goluzina, A.A. Lodkin, A.N. Podkorytov, *Problèmes d'Analyse Reéelle* (Cassini, Paris, 2010). French transl. of 2nd Russian edition (Nevskii Dialekt & BHV-Peterburg, St. Petersburg, 2004)
38. B. Makarov, A. Podkorytov, *Real Analysis: Measures, Integrals and Applications* (Springer, Berlin, 2013)
39. H.L. Montgomery, *Early Fourier Analysis*. The Sally Series, Pure and Applied Undergraduate Texts, vol. 22 (AMS, Providence, 2014)
40. I.P. Natanson, *Theory of Functions of a Real Variable* (Frederick Ungar Publishing, New York, 1955)
41. F. Natterer, *The Mathematics of Computerized Tomography*. Classics in Applied Mathematics, vol. 32. (SIAM, Philadelphia, 2001)
42. B. Opic, A. Kufner, *Hardy-Type Inequalities* (Longman, London, 1990)
43. J.N. Pandey, *The Hilbert Transform of Schwartz Distributions and Applications* (Wiley, New York, 1996)
44. M.A. Pinsky, *Introduction to Fourier Analysis and Wavelets*. Brooks/Cole Series in Advanced Mathematics (Brooks/Cole, Pacific Grove, 2002)
45. H. Reiter, J. D. Stegeman, *Classical Harmonic Analysis and Locally Compact Groups*, 2nd edn. London Math. Soc. Monographs. New Series, vol. 22 (The Clarendon Press, Oxford; University Press, New York, 2000)
46. R. Salem, *Essais sur les Séries Trigonometriques*, vol. 862 (Actual. Sci. et Industr., Paris, 1940)
47. G.E. Shilov, *Mathematical Analysis. A Special Course* (Pergamon Press, New York, 1965)

48. E.M. Stein, *Singular Integrals and Differentiability Properties of Functions* (Princeton Univ. Press, Princeton, 1970)
49. E.M. Stein, *Harmonic Analysis: Real-Variable Methods, Orthogonality, and Oscillatory Integrals* (Princeton Univ. Press, Princeton, 1993)
50. E.M. Stein, R. Shakarchi, *Functional Analysis: Introduction to Further Topics in Analysis* (Princeton Univ. Press, Princeton and Oxford, 2011)
51. E.M. Stein, G. Weiss, *Introduction to Fourier Analysis on Euclidean Spaces* (Princeton Univ. Press, Princeton, 1971)
52. A.E. Taylor, *General Theory of Functions and Integration* (Blaisdell Publ., New York, 1965)
53. A.F. Timan, *Theory of Approximation of Functions of a Real Variable* (Fizmatgiz, Moscow, 1960) (Russian). English transl.: Pergamon Press, MacMillan, N.Y. 1963 (there exist several later Dover publications of this translation)
54. E.C. Titchmarsh, *Introduction to the Theory of Fourier Integrals* (Oxford, 1937)
55. R.M. Trigub, E.S. Belinsky, *Fourier Analysis and Approximation of Functions* (Kluwer, Dordrecht, 2004)
56. N. Wiener, *The Fourier Integral and Certain of its Applications*(Dover Publ., New York, 1932)
57. A. Zygmund, *Trigonometric Series*, vol. I, II (Cambridge Univ. Press, Cambridge, 1968)

Papers

58. W. Abu-Shammala, A. Torchinsky, The atomic decomposition in $L^1(\mathbf{R^n})$. Proc. Am. Math. Soc. **135**, 2839–2843 (2007)
59. W. Abu-Shammala, A. Torchinsky, Spaces between H^1 and L^1. Proc. Am. Math. Soc. **136**, 1743–1748 (2008)
60. R. Akylzhanov, E. Liflyand, M. Ruzhansky, Re-expansions on compact Lie groups. Anal. Math. Phys. **10**, (2020) Paper No. 33, 20 pp.
61. R.A. Aliev, A.F. Amrahova, On the summability of the discrete Hilbert transform. Ural Math. J. **4**, 6–12 (2018)
62. K. Andersen, Weighted norm inequalities for Hilbert transforms and conjugate functions of even and odd functions. Proc. Am. Math. Soc. **56**(1), 99–107 (1976)
63. R. Askey, S. Wainger, Integrability theorems for Fourier series. Duke Math. J. **33**, 223–228 (1966)
64. B. Aubertin, J.J.F. Fournier, Integrability theorems for trigonometric series. Stud. Math. **107**, 33–59 (1993)
65. K.I. Babenko, An inequality in the theory of Fourier integrals. Izv. Akad. Nauk SSSR Ser. Mat. **25**, 531–542 (1961) (Russian)
66. S. Baron, E. Liflyand, U. Stadtmüller, The Fourier integrals of functions of bounded variation. Ann. Univ. Sci. Budapest. Eötvös Sect. Math. **49**, 43–52 (2006)
67. W. Beckner, Inequalities in Fourier analysis. Ann. Math. **102**, 159–182 (1975)
68. E.S. Belinsky, On asymptotic behavior of integral norms of trigonometric polynomials. Metric questions of the theory of functions and mappings. Nauk. Dumka Kiev **6**, 15–24 (1975) (Russian)
69. E. S. Belinsky, Application of the Fourier transform to summability of Fourier series. Sib. Mat. Zh. **XVIII**, 497–511 (1977) (Russian). English transl. in Siberian Math. J. **18**, 353–363 (1977)
70. E. Belinsky, E. Liflyand, R. Trigub, The Banach algebra A^* and its properties. J. Fourier Anal. Appl. **3**, 103–129 (1997)
71. J.J. Benedetto, H.P. Heinig, Weighted Fourier inequalities: new proofs and generalizations. J. Fourier Anal. Appl. **9**, 1–37 (2003)
72. O.V. Besov, Hörmander's theorem on Fourier multipliers. Trudy Mat. Inst. Steklov **173**, 164–180 (1986) (Russian). English transl. in Proc. Steklov Inst. Math. **4**, 4–14 (1987)

73. A. Beurling, On the spectral synthesis of bounded functions. Acta Math. **81**, 225–238 (1949)
74. R.P. Boas, Absolute convergence and integrability of trigonometric series. J. Ration. Mech. Anal. **5**, 621–632 (1956)
75. R.P. Boas, The integrability class of the sine transform of a monotonic function. Stud. Math. **XLIV**, 365–369 (1972)
76. S.V. Bochkarev, On a problem of Zygmund. Izv. Akad. Nauk SSSR, Ser. Mat. **37**, 630–638 (1973) (Russian). English transl. in Math. USSR Izvestija **37**, 629–637 (1973)
77. S. Bochner, A theorem on Fourier-Stieltjes integrals. Bull. Am. Math. Soc. **40**, 271–278 (1934)
78. D. Borwein, Linear functionals connected with strong Cesáro summability. J. Lond. Math. Soc. **40**, 628–634 (1965)
79. M. Bownik, Boundedness of operators on Hardy spaces via atomic decompositions. Proc. Am. Math. Soc. **133**, 3535–3542 (2005)
80. S. Boza, M.J. Carro, Discrete Hardy spaces. Stud. Math. **129**, 31–50 (1998)
81. J. S. Bradley, Hardy inequalities with mixed norms. Can. Math. Bull. **21**, 405–408 (1978)
82. M. Buntinas, N. Tanović-Miller, in *New Integrability and L^1-convergence Classes for Even Trigonometric Series II*, ed. by J. Szabados, K Tandori. Approximation Theory. Colloq. Math. Soc. János Bolyai, vol. 58 (North-Holland, Amsterdam, 1991), pp. 103–125
83. P.L. Butzer, The Hausdorff-Young theorems of Fourier Analysis and their impact. J. Fourier Anal. Appl. **1**, 113–130 (1994)
84. L. Carleson, On convergence and growth of partial sums of Fourier series. Acta Math. **116**, 135–157 (1966)
85. W. Cauer, The Poisson integral for functions with positive real part. Bull. Am. Math. Soc. **38**, 713–717 (1932)
86. J. Chen, D. Fan, S. Wang, Hausdorff operators on Euclidean spaces. Appl. Math. J. Chin. Univ. (Ser. B) (4) **28**, 548–564 (2014)
87. R. Coifman, G. Weiss, Extensions of Hardy spaces and their use in analysis. Bull. Am. Math. Soc. **83**, 569–645 (1977)
88. A. Cordoba, La formule sommatoire de Poisson. C.R. Acad. Paris, Ser. I **306**, 373–376 (1988)
89. H. Cramér, On the representation of functions by certain Fourier integrals. Trans. Am. Math. Soc. **46**, 190–201 (1939)
90. G. Dafni, E. Liflyand, A local Hilbert transform, Hardy's inequality and molecular characterization of Goldberg's local Hardy space. Complex Anal. Synerg. **5**(1), (2019) Paper No. 10, 9 pp.
91. S. Demir, A new inequality for the Hilbert transform (2020). arXiv: 2009.05822v1
92. A.G. Domínguez, The representation of functions by Fourier integrals. Duke Math. J. **6**, 246–255 (1940)
93. M. Dyachenko, E. Nursultanov, S. Tikhonov, Hardy-type theorems on Fourier transforms revised. J. Math. Anal. Appl. **467**, 171–184 (2018)
94. M.I. Dyachenko, S. Tikhonov, Convergence of trigonometric series with general monotone coefficients. C.R. Math. Acad. Sci. Paris **345**, 123–126 (2007)
95. R.E. Edwards, Criteria for Fourier transforms. J. Aust. Math. Soc. **7**, 239–246 (1967)
96. D. Faifman, A characterization of Fourier transform by Poisson summation formula. C. R. Acad. Paris Ser. I **348**, 407–410 (2010)
97. Ch. Fefferman, Characterizations of bounded mean oscillation. Bull. Am. Math. Soc. **77**, 587–588 (1971)
98. Ch. Fefferman, Pointwise convergence of Fourier series. Ann. Math. (2) **98**, 551–571 (1973)
99. Ch. Fefferman, E.M. Stein, H^p spaces of several variables. Acta Math. **129**, 137–193 (1972)
100. H.G. Feichtinger, A characterization of Wiener's algebra on locally compact groups. Arch. Math. (Basel) **29**, 136–140 (1977)
101. H.G. Feichtinger, Wiener amalgams over Euclidean spaces and some of their applications, in *Function Spaces (Edwardsville, IL, 1990)*. Lect. Notes Pure Appl. Math., vol. 136 (Dekker, New York, 1992), pp. 123–137

102. H.G. Feichtinger, F. Weisz, Herz spaces and summability of Fourier transforms. Math. Nachr. **281**, 1–16 (2008)
103. T.M. Flett, Some theorems on odd and even functions. Proc. Lond. Math. Soc. (3) **8**, 135–148 (1958)
104. T.M. Flett, Some elementary inequalities for integrals with applications to Fourier transforms. Proc. Lond. Math. Soc. (3) **29**, 538–556 (1974)
105. G.A. Fomin, A Class of trigonometric series. Matem. Zametki **23**, 213–222 (1978) (Russian). English transl. in Math. Notes **23**, 117–123 (1978)
106. J.J.F. Fournier, J. Stewart, Amalgams of L^p and ℓ^q. Bull. Am. Math. Soc. **13**, 1–21 (1985)
107. S. Fridli, Integrability and L^1-convergence of trigonometric and Walsh series. Ann. Univ. Sci. Budapest Sect. Comp. **16**, 149–172 (1996)
108. S. Fridli, On the L_1-convergence of Fourier series. Stud. Math. **125**(2), 161–174 (1997)
109. D.V. Giang, F. Móricz, On the integrability of trigonometric series. Anal. Math. **18**, 15–23 (1992)
110. D.V. Giang, F. Móricz, Multipliers of Fourier transforms and series on L^1. Arch. Math. **62**, 230–238 (1994)
111. D.V. Giang, F. Móricz, On the L^1-convergence of Fourier transforms. J. Aust. Math. Soc. (Ser. A) **60**, 495–420 (1996)
112. A. Gogatishvili, V.D. Stepanov, Reduction theorems for weighted integral inequalities on the cone of monotone functions. Uspekhi Mat. Nauk **68**, 3–68 (2013) (Russian). English transl. in Russ. Math. Surv. **68**, 597–664 (2013)
113. D. Goldberg, A local version of real Hardy spaces. Duke Math. J. **46**, 27–42 (1979)
114. R.R. Goldberg, Restrictions of Fourier transforms and extension of Fourier sequences. J. Approx. Theory **3**, 149–155 (1970)
115. D. Gorbachev, S. Tikhonov, Moduli of smoothness and growth properties of Fourier transforms: Two-sided estimates. J. Approx. Theory **164**, 1283–1312 (2012)
116. G.H. Hardy, J.E. Littlewood, Some properties of fractional integrals, I, II. Math. Z. **27**, 565–606 (1928), and **34**, 403–439 (1932)
117. G.H. Hardy, J.E. Littlewood, Some more theorems concerning Fourier series and Fourier power series. Duke Math. J. **2**, 354–382 (1936)
118. F. Hausdorff, Momentprobleme für ein endliches Intervall. Math. Zeitschrift **16**, 220–248 (1923)
119. C. Heil, in *An Introduction to Weighted Wiener Amalgams*, ed. by M. Krishna, R. Radha, S. Thangavelu. Wavelets and Their Applications (Chennai, 2002) (Allied Publishers, New Delhi, 2003), pp. 183–216
120. C.S. Herz, Lipschitz spaces and Bernstein's theorem on absolutely convergent Fourier transforms. J. Math. Mech. **18**, 283–323 (1968)
121. I.I. Hirschman, Jr., On multiplier transformations, I. Duke Math. J. **26**, 221–242 (1959); II, ibid **28**, 45–56 (1961)
122. F. Holland, Harmonic analysis on amalgams of L^p and l^q. J. Lond. Math. Soc. (2) **10**, 295–305 (1975)
123. R.A. Hunt, On the convergence of Fourier series, in *Orthogonal Expansions and Their Continuous Analogues (Proc. Conf., Edwardsville, Ill., 1967)* (Southern Illinois Univ. Press, Carbondale, 1968), pp. 235–255
124. R.A. Hunt, B. Muckenhoupt, R.L. Wheeden, Weighted norm inequalities for the conjugate function and Hilbert transform. Trans. Am. Math. Soc. **176**, 227–251 (1973)
125. S.I. Izumi, T. Tsuchikura, Absolute convergence of trigonometric expansions. Tôhoku Math. J. **7**, 243–251 (1955)
126. R.L. Johnson, C.R. Warner, The convolution algebra $H^1(R)$. J. Funct. Spaces Appl. **8**, 167–179 (2010)
127. J.-P. Kahane, Stylianos Pichorides. J. Geom. Anal. **3**, 533–542 (1993)
128. J.-P. Kahane, Y. Katznelson, Sur les ensembles de divergence des séries trigonométriques. Stud. Math. **26**, 305–306 (1966)

129. G. Karagulyan, On exceptional sets of the Hilbert transform. Real Anal. Exch. **42**, 311–327 (2017)
130. H. Kober, A note on Hilbert's operator. Bull. Am. Math. Soc. **48**, 421–426 (1942)
131. H. Kober, A note on Hilbert transforms. Q. J. Math. Oxford Ser. **14**, 49–54 (1943)
132. A. N. Kolmogorov, Sur l'ordre de grandeur des coefficients de la série de Fourier-Lebesgue. Bull. Acad. Polon. 83–86 (1923)
133. A.N. Kolmogorov, Une série de Fourier-Lebesgue divergente partout. C. R. Acad. Sci. Paris **183**, 1327–1328 (1926)
134. Yu. Kolomoitsev, E. Liflyand, Absolute convergence of multiple Fourier integrals. Stud. Math. **214**, 17–35 (2013)
135. Yu. Kolomoitsev, E. Liflyand, On weighted conditions for the absolute convergence of Fourier integrals. J. Math. Anal. Appl. **456**, 163–176 (2017)
136. J.C. Kuang, Generalized Hausdorff operators on weighted Morrey-Herz spaces. Acta Math. Sin. (Chin. Ser.) **55**, 895–902 (2012) (Chinese; Chinese, English summaries)
137. J.C. Kuang, Generalized Hausdorff operators on weighted Herz spaces. Mat. Vesnik **66**, 19–32 (2014)
138. M. Lacey, Chr. Thiele, A proof of boundedness of the Carleson operator. Math. Res. Lett. **7**, 361–370 (2000)
139. P. Lefèvre, A short direct proof of the discrete Hardy inequality. Arch. Math. **114**, 195–198 (2020)
140. A.K. Lerner, E. Liflyand, Interpolation properties of a scale of spaces. Collect. Math. **54**, 153–161 (2003)
141. E. Liflyand, On asymptotics of Fourier transform of functions from certain classes. Anal. Math. **19**, 151–168 (1993)
142. E. Liflyand, A family of function spaces and multipliers. Israel Math. Conf. Proc. **13**, 141–149 (1999)
143. E. Liflyand, On quasi-monotone functions and sequences. Comput. Methods Funct. Theory **1**, 345–352 (2001)
144. E. Liflyand, Lebesgue constants of multiple Fourier series. Online J. Anal. Combin. **1**, 112 p. (2006)
145. E. Liflyand, On absolute convergence of Fourier integrals. Real Anal. Exch. **36**, 353–360 (2010/2011)
146. E. Liflyand, Fourier transform versus Hilbert transform. Ukr. Math. Bull. **9**, 209–218 (2012). Also published in J. Math. Sci. **187**, 49–56 (2012)
147. E. Liflyand, Fourier transforms on an amalgam type space. Monatsh. Math. **172**, 345–355 (2013)
148. E. Liflyand, Hausdorff operators on Hardy spaces. Eurasian Math. J. **4**, 101–141 (2013)
149. E. Liflyand, On Fourier re-expansions. J. Fourier Anal. Appl. **20**, 934–946 (2014)
150. E. Liflyand, Interaction between the Fourier transform and the Hilbert transform. Acta Commun. Uni. Tartu. Math. **18**, 19–32 (2014)
151. E. Liflyand, Integrability spaces for the Fourier transform of a function of bounded variation. J. Math. Anal. Appl. **436**, 1082–1101 (2016)
152. E. Liflyand, Weighted estimates for the discrete Hilbert transform, in *Methods of Fourier Analysis and Approximation Theory*, Appl. Numer. Harmon. Anal. (Birkhäuser/Springer, 2016), pp. 59–69
153. E. Liflyand, The Fourier transform of a function of bounded variation: symmetry and asymmetry. J. Fourier Anal. Appl. **24**, 525–544 (2018)
154. E. Liflyand, Salem conditions in the nonperiodic case. Mat. Zametki **104**, 447–453 (2018) (Russian). English transl. in Math. Notes **104**, 437–442 (2018)
155. E. Liflyand, Hardy type inequalities in the category of Hausdorff operators, in *Modern Methods in Operator Theory and Harmonic Analysis. OTHA 2018. Springer Proceedings in Mathematics & Statistics*, ed. by A. Karapetyants, V. Kravchenko, E. Liflyand, vol. 291 (Springer, Berlin, 2019), pp. 81–91
156. E. Liflyand, Ball's lemma as an exercise. Chebyshevskii Sbornik **54** (2021)

157. E. Liflyand, L^1 convergence of Fourier transforms. Anal. Math. Phys. **11** (2021)
158. E. Liflyand, A. Miyachi, Boundedness of the Hausdorff operators in H^p spaces, $0 < p < 1$. Stud. Math. **194**(3), 279–292 (2009)
159. E. Liflyand, F. Móricz, The Hausdorff operator is bounded on the real Hardy space $H^1(\mathbb{R})$. Proc. Am. Math. Soc. **128**, 1391–1396 (2000)
160. E. Liflyand, E. Ournycheva, Two spaces conditions for integrability of the Fourier transform. Analysis (Munich) **28**, 429–443 (2008)
161. E. Liflyand, S. Samko, R. Trigub, The Wiener algebra of absolutely convergent Fourier integrals: an overview. Anal. Math. Phys. **2**, 1–68 (2012)
162. E. Liflyand, S. Tikhonov, Extended solution of Boas' conjecture on Fourier transforms. C. R. Math. Acad. Sci. Paris **346**, 1137–1142 (2008)
163. E. Liflyand, S. Tikhonov, The Fourier Transforms of General Monotone Functions, in *Analysis and Mathematical Physics*. Trends in Mathematics (Birkhäuser, Boston, 2009), pp. 373–391
164. E. Liflyand, S. Tikhonov, Weighted Paley–Wiener theorem on the Hilbert transform. C.R. Acad. Sci. Paris, Ser. I **348**, 1253–1258 (2010)
165. E. Liflyand, S. Tikhonov, A concept of general monotonicity and applications. Math. Nachr. **284**, 1083–1098 (2011)
166. E. Liflyand, S. Tikhonov, M. Zeltser, Extending tests for convergence of number series. J. Math. Anal. Appl. **377**, 194–206 (2011)
167. E. Liflyand, R. Trigub, Conditions for the absolute convergence of Fourier integrals. J. Approx. Theory **163**, 438–459 (2011)
168. E. Liflyand, R. Trigub, Wiener algebras and trigonometric series in a coordinated fashion. Constructive Approx. **54** (2021). arXiv:1910.02777
169. Y.-W. Liu, Hilbert transform and applications. *Fourier Transform Applications*, in (InTech, 2012), pp. 291–300
170. V. Matsaev, M. Sodin, Distribution of Hilbert transforms of measures. Geom. Funct. Anal. **10**, 160–184 (2000)
171. A. Miyachi, Weak factorization of distributions in H^p spaces. Pac. J. Math. **115**, 165–175 (1984)
172. F. Móricz, On the integrability and L^1-convergence of complex trigonometric series. Proc. Am. Math. Soc. **113**, 53–64 (1991)
173. R. Nevanlinna, Asymptotische Entwicklungen beschränkter Funktionen und das Stieltjessche Momentenproblem. Ann. Acad. Sci. Fenn. Ser. A **18**, 1–53 (1922)
174. L. Nirenberg, On elliptic partial differential equations. Ann. Scuola Norm. Sup. Pisa (3) **13**, 115–162 (1959)
175. R. Oberlin, A. Seeger, T. Tao, Chr. Thiele, J. Wright, A variation norm Carleson theorem. J. Eur. Math. Soc. **14**, 421–464 (2012)
176. A.M. Olevskii, Singularities of Carleman type for complete orthonormal systems. Sibirsk. Mat. Z. **8**, 807–826 (1967) (Russian)
177. R.E.A.C. Paley, N. Wiener, Notes on the theory and application of Fourier transform. Note II, Trans. Am. Math. Soc. **35**, 354–355 (1933)
178. M.A. Pinsky, Fejér asymptotics and the Hilbert transform, in *Harmonic analysis at Mount Holyoke (South Hadley, MA, 2001)*. Contemp. Math., vol. 320 (Amer. Math. Soc., Providence, 2003), pp. 333–340
179. J. Pfleger, Über Reihen mit r-fach monoton abnehmenden Gliedern. Monatsh. Math. Phys. **41**, 191–200 (1934)
180. A. Plessner, Eine Kennzeichnung der totalstetigen Funktionen. J. Reine Angew. Math. **160**, 26–32 (1929)
181. P.G. Rooney, On the representation of sequences as Fourier coefficients. Proc. Am. Math. Soc. **11**, 762–768 (1960)
182. R. Ryan, Fourier transforms of certain classes of integrable functions. Trans. Am. Math. Soc. **105**, 102–111 (1962)

183. D. Ryabogin, B. Rubin, Singular integral operators generated by wavelet transforms. Integr. Equ. Oper. Theory **35**, 105–117 (1999)

184. C. Sadosky, R.L. Wheeden, Some weighted norm inequalities for the Fourier transform of functions with vanishing moments. Trans. Am. Math. Soc. **300**, 521–533 (1987)

185. Y. Sagher, Integrability conditions for the Fourier transform. J. Math. Anal. Appl. **54**, 151–156 (1976)

186. S. Szidon, Reihentheoretische Sätze und ihre Anwendungen in der Theorie der Fourierischen Reihen. Math. Z. **10**, 121–127 (1921)

187. S. Szidon, Hinreichende Bedingungen für den Fourier-Charakter einer trigonometrischen Reihe. J. Lond. Math. Soc. **14**, 158–160 (1939)

188. E.M. Stein, Note on the class $L \log L$. Stud. Math. **32**, 305–310 (1969)

189. C. Sweezy, Subspaces of $L^1(\mathbb{R}^d)$. Proc. Am. Math. Soc. **132**, 3599–3606 (2004)

190. M. Taibleson, Fourier coefficients of functions of bounded variation. Proc. Am. Math. Soc. **18**, 766 (1967)

191. S.A. Telyakovskii, An estimate, useful in problems of approximation theory, of the norm of a function by means of its Fourier coefficients. Trudy Mat. Inst. Steklova **109**, 65–97 (1971) (Russian). English transl. in Proc. Steklov Math. Inst. **109**, 73–109 (1971)

192. S.A. Telyakovskii, Concerning a sufficient condition of Sidon for the integrability of trigonometric series. Mat. Zametki **14**, 317–328 (1973) (Russian). English transl. in Math. Notes **14**, 742–748 (1973)

193. S. Tikhonov, Trigonometric series with general monotone coefficients. J. Math. Anal. Appl. **326**, 721–735 (2007)

194. S. Tikhonov, Best approximation and moduli of smoothness: computation and equivalence theorems. J. Approx. Theory, **153**(1), 19–39 (2008)

195. R.M. Trigub, Absolute convergence of Fourier integrals, summability of Fourier series, and polynomial approximation of functions on the torus. Izv. Akad. Nauk SSSR, Ser.Mat. **44**, 1378–1408 (1980) (Russian). English translation in Math. USSR Izv. **17**, 567–593 (1981)

196. R.M. Trigub, A generalization of the Euler-Maclaurin formula. Mat. Zametki **61**, 312–316 (1997) (Russian). English transl. in Math. Notes **61**, 253–257 (1997)

197. F. Weisz, Herz spaces and restricted summability of Fourier transforms and Fourier series. J. Math. Anal. Appl. **344**, 42–54 (2008)

198. N. Wiener, On the representation of functions by trigonometric integrals. Math. Z. **24**, 575–616 (1926)

199. J. Xiao, L^p and BMO bounds of weighted Hardy–Littlewood averages. J. Math. Anal. Appl. **262**, 660–666 (2001)

200. W.H. Young, On the Fourier series of bounded functions. Proc. Lond. Math. Soc.(2) **12**, 41–70 (1913)

201. P. Zorin-Kranich, Maximal polynomial modulations of singular integrals (2018). arXiv: 1711.03524v5

202. A. Zygmund, Some points in the theory of trigonometric and power series. Trans. Am. Math. Soc. **36**, 586–617 (1934)

Index

Printed in the United States
by Baker & Taylor Publisher Services